Best Wine Buys
At Under-A-Fiver 2000

Best Wine Buys
At Under-A-Fiver 2000

Ned Halley

foulsham
LONDON • NEW YORK • TORONTO • SYDNEY

foulsham

The Publishing House
Bennetts Close, Cippenham, Berkshire SL1 5AP, England

ISBN 0-572-02559-9

Text copyright © 2000 Ned Halley

Series, format and layout design © 2000 W. Foulsham & Co. Ltd

Cover photograph © Anthony Blake Photo Library.

Printed in Great Britain by Cox & Wyman Ltd, Reading.

CONTENTS

A personal note

All the wines mentioned in this book are recommended, and some are more recommended than others. The **V** symbol signifies exceptional value for money. The ● symbol denotes a wine either of special quality or particular individuality – or even both. Every wine appended with a (VQ) symbol should be tried at the earliest possible opportunity.

This is the point of *Best Wine Buys Under a Fiver*. The wines described in these pages are the ones I have picked out from around two thousand tasted in 1998–99. Qualifying wines have met simple criteria. Apart from costing under £5, they have to give genuine pleasure. I have not excluded many wines on the basis that they are bad or faulty. Quality control by retailers ensures that standards are reassuringly high, even at the bottom of the price scale.

But I have left out hundreds of wines on the basis that I can't imagine anyone wanting to drink them. Overpriced wines are the prime offenders. There are plenty at £4 plus that wouldn't be good value even at two quid less. And there have been many wines that fail to pass the 'so what' test. You know, as in answering the question: 'So what do you like about this wine?'

If I have not been able to write down what that is, I cannot recommend the wine. I have to like a wine for a reason, and to be able to articulate that reason. Otherwise, I would be in no position to write a book such as this.

I do hope you will enjoy reading and using *Best Wine Buys*. I must, of course, apologise in advance for the inevitable fact that some of the wines mentioned will have been discontinued or displaced with a new vintage or increased in price by the time you are reading this. And I had better show contrition, too, for any disagreements readers may have with me over the comments I have made about individual wines.

Taste is personal in all things, and more so than most when it comes to wine. But I hope the impressions I have given of the hundreds of wines recommended in this guide will tempt to you to try some new flavours, and perhaps even to trade up to a few of the small selections I have made of wines above the £5 limit.

If *Best Wine Buys* arouses your curiosity about the wider world of wine, and encourages you to experiment, then it will have achieved its objective.

Where do the best wines come from?

This question once had a straightforward answer: France. It became a bit more complicated when Australia started making good wine in the early 1980s. Now the question is hardly answerable at all. South American wines have appeared as if from nowhere. South Africa is back on the map. Italy, Portugal and Spain now produce top-quality, bargain-priced wines with names unheard-of just a few years ago. Bulgaria had a good spell for a while but has waned. Hungary, invigorated by foreign investment, is improving fast. So is Austria. Germany has seen worldwide sales of Liebfraumilch plummet, but continues, as it always has, to produce some of the best white wines in the world.

France, meanwhile, strives to maintain pole position. The new French boom is in wines from the Midi, the deep south. Here, as in Australia, they are building hi-tech production centres that can make fresh and lively wines even in the hottest climes of the region. Some of these wines bear a startling resemblance to the upstart competition from the southern hemisphere.

Among less expensive wines, the theme is a varietal one. In other words, the wines are labelled with the name of the grape variety from which they are made. The main selling point for most of these wines is no longer the country of origin, but the grape of origin. We tend not to express preferences for French, Italian or Californian wine. Instead we specify Chardonnay, Cabernet Sauvignon or Sauvignon Blanc.

It makes sense, because the characteristics of various grape varieties do so much more to identify wines than a country of origin can ever do. A bottle of white wine labelled Chardonnay can reasonably be counted on to deliver that distinctive peachy or pineappley smell and soft, unctuous, apple flavours. A Sauvignon Blanc should evoke gooseberries, green fruit and crisp freshness, and so on.

But there is a drawback. Wines from a given grape variety grown in vineyards with similar climatic conditions and made by the same modern methods can taste uncomfortably alike. In the age of 'globalisation' we just have to accept that this is the way things are. But it means enthusiasts in search of genuine variety must expect to work a little harder, as time goes by, to find genuinely individual wines.

As to the best sources of wines under a fiver, it will take only a brief perusal of this book to reveal that some parts of the world appear to offer better value than others. The classic regions of France – Alsace, Bordeaux, Burgundy – make few appearances, simply because their wines are now almost exclusively priced above the £5 threshold.

The vins de pays ('country wines') of France's less vaunted regions, on the other hand, appear on every page. So do the wines of Argentina, which unquestionably offers the best value, especially under £5, of any nation in the world.

In this book, you'll find wines made from 40 different named grape varieties grown in fewer than 20 different countries. It's the shape of things to come.

The state of the market

Britain drinks about a billion bottles of wine each year. It sounds a lot, but only comes to about half a bottle of wine per adult per week. Still, it's a start. Back in 1900, per capita wine consumption here was less than two bottles per year. This dismal level prevailed until the second half of the century, showing improvement only when foreign holidays became the norm, and large numbers of us suddenly discovered a taste for continental vices.

But long before the package-holiday era, Britain was nevertheless an important international market for wine. For centuries, keen imbibers here have been drinking more champagne, claret, Moselle and port than in any other country. So when cheaper wines started to flood in during the 1960s, our wine trade was already geared up to take advantage of the new boom in low-margin plonk.

Unfortunately for the long-established specialist merchants, the supermarkets have scooped most of the market, but Britain has continued to be, as it has always been, the best country in the world in which to buy wine. Not the cheapest – thanks to a long Puritan tradition of trying to repress alcohol consumption through usurious excise duties – but with by far the widest range of sources.

In spite of the high costs prevailing here, Britain drinks more good-quality imported wine than any other country in the world. We are the most valued export market for several of the leading wine-producing nations.

Thus the astounding choice in the shops. Supermarkets carry from 500 to 700 distinct brands. The wine department in Safeway or Tesco can be depended upon to take up more shelf space than bread or meat. But shopping for wine is a rather more taxing task than choosing a loaf or picking out something for the Sunday roast. Hence this new edition of *Best Wine Buys*, which I hope will provide a practical primer to what should be one of the more pleasurable aspects of shopping.

Supermarkets on top

In a nation with so many traditional wine merchants, from grand London purveyors with Royal Warrants to tiny specialist firms importing the wines of just one region of France, it comes as a shock to find that three-quarters of all the wine bought 'off-licence' (for drinking at home) in Britain comes from supermarkets.

But are supermarkets selling quantity at the expense of quality? Not a bit of it. These huge companies take quality very seriously – under £5 (accounting for nine out of every ten bottles sold) as much as over.

Tasting hundreds of supermarket wines every year as I do, I can honestly report finding few I could fairly describe as badly made or unpleasant to drink. But even making allowances for the jading effects of tasting scores of wines at one session, I must confess to one slightly troubling discovery. It is the striking homogeneity of an awful lot of wines. Far too many taste the same as each other.

Fair's fair, there is in some cases a simple explanation for this – because many supermarket own-brand wines *are* the same as each other. Big producers in every part of the world happily apportion their harvests between two, three or more British supermarket chains, bottling the wine in one giant operation with pauses merely for changing the labels.

And why not? The more outlets for the brilliant wines of Argentina – currently the world's clear winner for wine value – the better. One major winery in the Mendoza region supplies its wines to all but one of Britain's six biggest supermarkets, each of which sells them under its own label.

The sameness problem, however, isn't really about the enviable success of Argentine exporters. It's more about the uniformity of style. Supermarkets have convinced themselves that they know what their customers want. Most of the wines on the shelves conform not just to the prices dictated by the wine buyers, but to the styles they expect. Safeway, Sainsbury and Tesco all have a hand in making many of the wines they sell, demanding that the products of wineries in every part of the world conform to the styles customers are perceived to demand.

This can hardly be faulted. But wine shoppers who do experiment with different wines on a regular basis will find that there can be a depressing similarity in, say, Chardonnay from sources as diverse as Hungary, Chile, South Africa and the far south of France. The wine-taster's party trick of identifying the nationality of an everyday wine is getting to be a very difficult one to perform with any hope of success.

That said, there is little room for complaint about the supermarkets. There is a terrific choice, and commendable wines under a fiver proliferate. Thus the very extensive sections devoted to some of the supermarkets in the following pages.

High street combinations

When Britain's two biggest high street off-licence chains, Thresher and Victoria Wine, announced their 1999 amalgamation under the strikingly odd name 'First Quench', they proclaimed that their combined network of shops would give them a 15 per cent share of the take-home wine market. It sounds impressive enough until you consider that Tesco alone, through its 566 supermarkets, was already selling more wine than all 3,000 First Quench shops put together.

This puts the traditional off-licence shop firmly in perspective. Had the boom in wine-drinking over the last 20 years not taken place, it seems unlikely even big chains like Threshers would have survived at all in the face of supermarket competition. Whether the strategy of merging into huge nationwide chains will pay off remains to be seen.

In the meantime, it's only fair to point out that off-licences are in many ways much more attractive places in which to shop for wine than supermarkets. Prices are quite definitely competitive (it is a myth that supermarket wines are cheap) and if you're prepared to buy a dozen bottles of wine at a time, the likes of Oddbins, Thresher-Victoria Wine, Fullers and Wine Cellars will always give you 10 per cent off.

Off-licences are forever running special promotions on their wines, too. In the struggle to compete with supermarkets, they have 'loyalty' schemes which outdo store 'points' cards by a mile for value. If you are even an occasional customer at a local branch of a major off-licence chain, don't hesitate to ask about any club-style arrangements they might have.

'Fine wine' merchants

If the nationwide high street off-licence chains have been squeezed by the hypermarkets, how much more so for the independently owned specialist wine merchants? It's a testimony to the enterprise of centuries-old firms such as Berry Bros & Rudd in London that they can continue to prosper in the age of the retail multiple – but they do.

Grand old wine merchants can be daunting. It is quite understandable that those proudly proclaiming themselves 'purveyors of fine wines to the Royal Household' and the like should deter all but the boldest enthusiast in search of something a bit more interesting than this month's special offer at the supermarket.

It's true that some firms specialising in classed-growth claret and burgundy don't tend to offer much in the way of wines under £5, but this is not a universal truth. Berry Bros is not included in this book, but several other long-established specialist wine merchants are – for the simple reason that they have plenty of excellent wines within the price limit.

What's more, independent merchants are much better placed to offer wines from smaller producers. As can be imagined, most of the world's best wines are made by individual farmer-winemakers rather than by industrial-scale corporations. But supermarkets and shop-chains with hundreds of outlets to stock are not interested in wines made in small quantities. So if you have an adventurous taste for individualistic wines, look to the independent firms, who make a speciality of seeking out interesting small producers.

All the independent firms featured in this book offer nationwide delivery by mail order. Each publishes a detailed list at least once a year, providing helpful descriptions of all their wines along with prices, delivery details and so on. If you get on to their mailing lists, they will send you occasional special offers, news of events such as tastings, the occasional 'bin-end' sale. It's all a bit like joining a club – though many temptations to try better and better wines will be put in your way.

The price of wine

How do retailers price their wines? Some bottles seem inexplicably cheap, others unjustifiably expensive. But there is often a simple explanation. Supermarkets work to price points. In wine, these are £2.99, £3.49, £3.99 and so on. You'll find only a few bottles priced anywhere between these 50p spacings. A wine that wouldn't be profitable at £2.99 but would be at, say, £3.11 is quite likely to be priced at £3.49 in the hope that shoppers won't be wise to the fact that it is relatively poor value.

It's true that there are some wines on supermarket shelves priced at £3.29, £3.79 etc. but these price points occur with suspicious irregularity, and suggest that an awful lot of wines are being pushed the greater distance towards the next 49 and 99 pence points.

Price can be a poor guide to quality even at the best of times. The only means by which any of us can determine a wine's value is on personal taste. The ideal bottle is one you like very much and would buy again at the same price without demur.

But just for curiosity's sake, it's fun to know what the wine itself actually costs, and what the retailer is making on it. This is how the costs break down in a French wine costing £4.49 at a mythical supermarket we'll call Asway. This is a slightly unusual purchase by a supermarket, because the wine is being bought direct from the vineyard where it was made. Usually, retail multiples buy their wines by a less strenuous method, from agents and distributors in the UK.

Price paid by Asway to the producer for the bottled wine	£1.50
Transport and insurance to UK	£0.28
Excise duty	£1.12
Cost to Asway	£2.90
Asway's routine mark-up at 30 per cent	£0.87
VAT at 17.5 per cent on marked-up price	£0.66
Provisional shelf price	£4.43
Adjustment in price/VAT to price point	£0.06
Actual price in Asway	£4.49

The largest share of the money appears to go to the producer. But from his £1.50 he must pay the cost of growing and harvesting the grapes, pressing them, fermenting the juice, clarifying and treating the wine. Then he must bottle, cork, encapsulate, label and pack the wine into cartons. If his margin after these direct costs is 50 pence, he's doing well.

The prime profiteer, however, is not the supermarket, even though it makes a healthy 92 pence in mark-up. It is the Chancellor who does best, by miles. Excise duty and VAT are two of the cheapest taxes to collect – less than 1 per cent of revenue raised – and from this single bottle of wine, the Treasury trousers a princely £1.79.

Travellers to wine-producing countries are always thrilled to find that by taking their own bottles, jugs or giant demi-johns to rustic vineyards offering wine from the cask, they can buy drinkable stuff for as little as 50 pence a litre. What too few travellers appreciate is that, for the wine itself, that's about what the supermarkets are paying for it. When enjoying your sub-£5 bottle of wine, it is interesting to reflect on the economic reality known as 'added value' – which dictates that the worthiest person in the chain, the producer, has probably earned less than ten per cent of the price.

Cross-Channel shopping

Is it really worth travelling across the Channel for the sole purpose of buying wine at cheaper prices than those prevailing in Britain? The short answer is yes. It's fun to visit France (or even, at a pinch, Belgium) and out of season it can be very cheap. High-speed ferries, catamarans and hovercraft can carry you to historic Channel ports in an hour or two for just a few pounds if you take advantage of the perpetual ticket promotions run in the national press. Discounted tickets through the Eurotunnel are never quite as cheap, but for train enthusiasts, the journey is a treat in itself.

True, there is no longer any duty-free shopping on board the ferries (there never was on the trains), but as passengers on P&O Stena, SeaFrance and Hoverspeed have been discovering since the old tax-perk was abolished on 30 June 1999, this has made no difference. You can still buy wine, spirits and other goodies on board the ships at prices that seem remarkably adjacent to those pertaining in the good old days of duty-free.

The reason for this curious continuity is that shipping companies are now buying their supplies duty-paid on the other side of the Channel – which in some cases is almost as cheap as buying those supplies on a duty-free basis used to be. After all, the duty on wine in France is only 2p per 75cl bottle. In Spain, Italy and elsewhere in southern Europe, there's no duty on wine at all. VAT in France, at 20.6 per cent, is appreciably higher than the prevailing rate of 17.5 per cent in Britain, but ferry operators are absorbing the cost.

In effect, the ships are in direct competition with the supermarkets and hideous British-owned 'wine warehouses' in the Channel ports. But ferry operators cannot hope, really, to compete with France's huge and powerful hypermarket companies. Auchan, Cora, Continent, Leclerc and other retail multiples have enormous buying power and of course very much larger premises in which to display their goods. And the car park of a major out-of-town *hypermarché* has rather more room in which to manoeuvre a groaning trolley than does the vehicle deck of a roll-on, roll-off ferry.

As explained in the feature on pages 18–19, wine prices are dramatically lower in French supermarkets than they are in their British counterparts. You can discover this most graphically by visiting Tesco's 'Vin Plus' branch in the Cité Europe shopping centre at the Pas de Calais (near the Eurotunnel terminal) or Sainsbury's drinks-department shop next to Calais' mega Auchan supermarket.

But it is infinitely more exciting to shop in the French *hypermarchés* themselves. Just about everything is cheaper – not just the wine, spirits and beers – and there is a choice of fresh and preserved foods in the larger superstores that puts our own supermarkets to shame. English is widely spoken, and bilingual signs are commonplace in the stores. And if you're using an appropriate credit card, paying the bill is no more complicated than it is in a supermarket at home.

There is no limit to the quantity or value of goods you can bring back from any EU member country, provided they are not intended for resale in the UK. British Customs and Excise long ago published guidelines as to what they consider are reasonable limits on drinks that can be deemed to be for 'personal consumption'. You can thus import, no questions asked, 90 litres of wine, 20 of fortified wine, 10 of spirits and 110 of beer. That's about as much as a couple travelling together (and therefore able to import twice the above quantities) could cram into a family car without threatening the well-being of its suspension.

Savings on beer and spirits are, if anything, even more dramatic than they are on wine. French duty on beer is 5p a pint, compared to 33p here. This means a case of beer typically costing £12–15 here can be had for under a fiver in Calais. It seems crazy. Similarly, a 70cl bottle of London gin or Scotch whisky costing £11–12 here is yours for £7–8 in France, where duty on spirits is half the £5.48 charged here and mark-ups are lower.

As if these differentials were not enough, the Channel ports also teem with good-value restaurants and hotels. Boulogne and Calais, Dieppe and Dunkirk bristle with venues where you can enjoy a 100-franc menu of a standard that would cost several times as much back home. And there are respectable hotels where a clean room with bath or shower en suite, plus croissants and excellent coffee for breakfast can be had for £25 all in.

Why are wines so much cheaper in France?

The price gaps between the big stores on either side of the Channel are by no means entirely accounted for by the difference between UK excise duty and VAT and French duty and VAT. So what's going on?

It's not that British supermarkets charge higher margins than their continental counterparts. True, the Office of Fair Trading's Competition Commissioner is currently investigating allegations that supermarkets are overcharging customers. But when it comes to wine, the differential between the same wine on sale in the Asda at Coventry and the Auchan at Calais arises from *how* those respective supermarkets apply their margins.

It goes like this. In France, retailers typically mark up wines at 30 per cent. In the UK, as it happens, retailers also mark up by around 30 per cent. The difference is that shops in Britain add the mark-up not to the basic value of the wine, but to the duty paid and delivered (DPD) price.

In the UK, every bottle of still wine, regardless of quality or price, comes with an excise duty and shipping cost of around £1.40. So a bottle of wine the retailer buys for £1 from the producer has a DPD price of £2.40. Marked up by 30 per cent for the retailer's margin, that becomes £3.12. Add VAT at 17.5 per cent and the actual retail price turns into £3.66.

In France, it's very different. The typical duty and shipping cost in the price of a bottle of French wine is 12p. So the DPD price of the £1 bottle is £1.12. Marked up by 30 per cent it becomes £1.46. Add French VAT at 20.6 per cent and the actual retail price is £1.76. That's nearly £2 lower than the price of the same bottle of wine on our side of the Channel.

The best buys

The question, 'Which is your favourite wine?' is a familiar one to anyone who writes on this subject. It's impossible to answer without several qualifications – red or white, sweet or dry, under £5 or over £5. But in the context of this book, it's relatively straightforward. Here are the wines from the following pages that I rate most highly for value. All score well for quality and individuality, but the rank order is really determined on the basis of the quality-value ratio.

Top ten red wines under £5

PORTADA
VINHO REGIONAL
ESTREMADURA

TINTO 1997
PRODUCE OF PORTUGAL
PRODUZIDO E ENGARRAFADO
POR: ENG. 154
2560 - PORTUGAL

11,5% vol   75cl e

1   Portada Red 1997 £3.99 Somerfield, Unwins, Wine Cellars

2   Caballo de Plata Bonarda/Barbera 1998 £3.49 Safeway

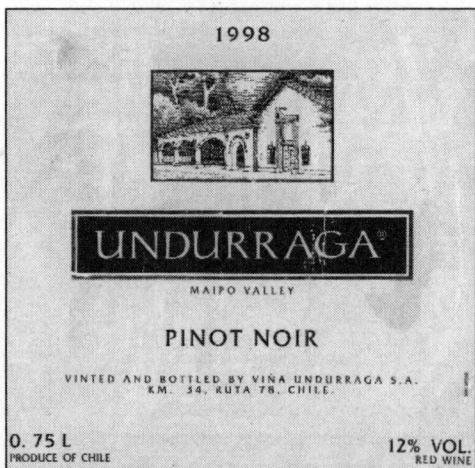

3   Undurraga Pinot Noir 1998 £4.99 Tesco

4   Piornos Vinho Regional Beiras 1997   £4.25 Adnams

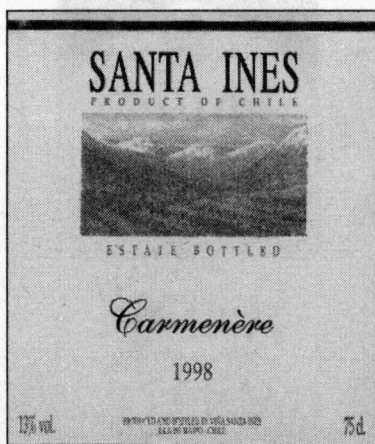

SANTA INES

PRODUCT OF CHILE

ESTATE BOTTLED

Carmenère

1998

13% vol.   75cl

5   Santa Ines Carmenère 1998   £4.49   Tesco

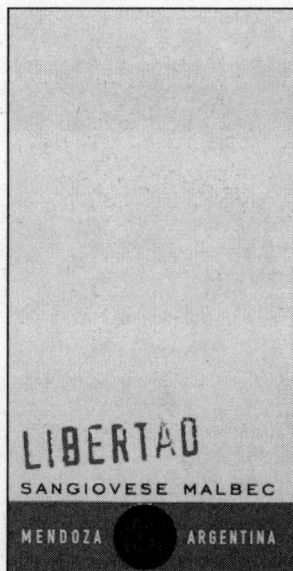

6   Libertad Malbec/Sangiovese 1998 £3.50 Bibendum (£3.79 First Quench/Fuller's)

7   Dão Dom Ferraz 1997 Booths, Budgens, First Quench, Tesco

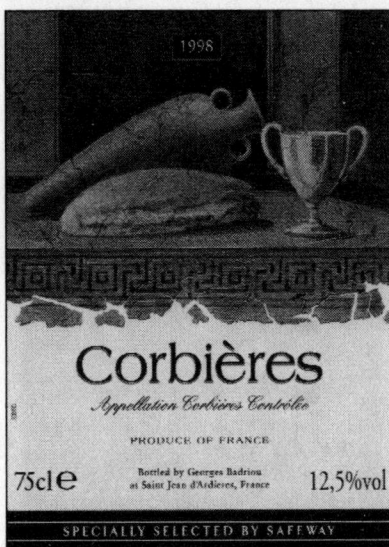

8 Safeway Corbières 1998 £3.19 Safeway

9 Promesa Tinto 1994 £3.85 Tanners

10   **Prosperity Red £4.99 Majestic**

Top ten white wines under £5

1   Denis Marchais Hand Picked Vouvray 1998   £4.99   Asda

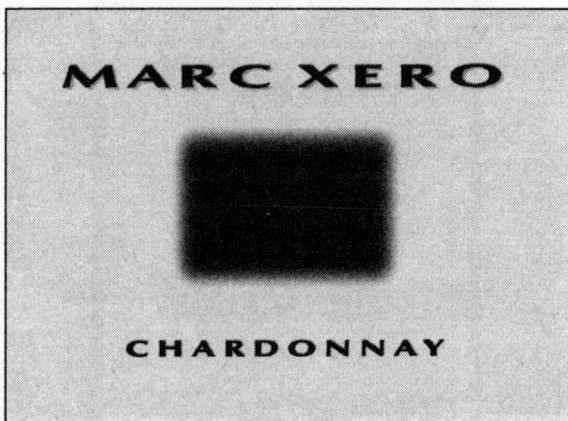

MARC XERO

CHARDONNAY

2   Marc Xero Chardonnay 1998 £4.99 Safeway/Sainsbury's/Somerfield/Tesco

MOSEL-SAAR-RUWER

1992er

Ockfener Bockstein

Riesling

KABINETT

Qualitätswein mit Prädikat
A. P. Nr. 3 561 107 034 93

Alc. 8.0 % by vol   Gutsabfüllung   **750 ml e**

Staatliche Weinbaudomäne Trier
D-54290 Trier

PRODUCE OF GERMANY

3   Ockfener Bockstein Riesling Kabinett, Jacobus, 1992 £4.99 Majestic

SAINSBURY'S

SICILIA

INDICAZIONE GEOGRAFICA TIPICA

White Wine

PRODUCE OF ITALY

ALCOHOL 11.5% BY VOLUME ℮75CL

Bottled for Sainsbury's Supermarkets Ltd. Stamford Street London SE1 9LL UK
by scarl Cantine Settesoli Menfi Italy

4   Sicilian White   £2.99   Sainsbury's

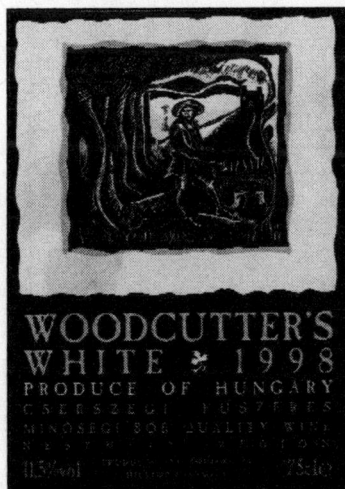

WOODCUTTER'S
WHITE ☙ 1998
PRODUCE OF HUNGARY

11.5% vol   75cl℮

5   Woodcutter's White 1998   £2.99   Safeway

6   Domaine de Raissac Viognier 1998 £4.99 Majestic

7   Old Penola Estate Botrytis Gewürztraminer (½ bottle) 1997 £4.99 Tesco

SAINSBURY'S
Special Cuvée

NIERSTEINER GUTES DOMTAL

Qualitätswein
Rheinhessen

A.P. Nr. 4 342 899 339 99

PRODUCE OF GERMANY

Alcohol 10% by volume

75 cl e

SELECTED BY SAINSBURY'S SUPERMARKETS LTD, STAMFORD STREET LONDON SE1 9LL UK
BOTTLED BY D-RP 342 899 SHIPPED BY ST. LAURENS WEINKELLEREI GmbH BINGEN GERMANY

8   Special Cuvée Niersteiner Gutes Domtal £2.99 Sainsburys

Produce of France

DOMAINE de RIEUX
1998

Vin de Pays
des Côtes de Gascogne

Mis en bouteille au Domaine

J. Grassa Fille et Fils

Domaine de Rieux
St-Amand-Eauze
32800 (Gers) France

11% vol.
75cl

9   Domaine de Rieux 1998 £4.25 Balls Brothers (£4.50 Tanners)

VIN D'ALSACE
APPELLATION ALSACE CONTRÔLÉE

BLANC DE BLANCS

1997

MIS EN BOUTEILLE PAR

CAVE DES VIGNERONS
A TURCKHEIM (HT-RHIN) FRANCE

PRODUCT OF FRANCE

12%vol. 750ml

10   Alsace Blanc de Blancs, Turckheim, 1997 £3.99 Somerfield

Making the most of it

There has always been a lot of nonsense talked about the correct way to serve wine. Red wine, we are told, should be opened and allowed to 'breathe' before pouring. White wine should be cool but not cold. Wine should never be drunk with soup, tomatoes or chocolate. You know the sort of thing.

It would all be simply laughable except that these daft conventions do make so many potential wine lovers nervous about the simple ritual of opening a bottle and sharing it around. Here is a short and opinionated guide to the received wisdom.

Breathing
Simply uncorking a wine for an hour or two before you serve it will make absolutely no difference whatsoever to the way it tastes. However, if you wish to warm up an icy bottle of red by placing it near (never on) a radiator or fire, do remove the cork first. As the wine warms, even very slightly, it expands.

Chambré-ing
One of the more pretentious terms in the wine vocabulary. The idea is that red wine should be at the same temperature as the room *(chambre)* you're going to drink it in. In fairness, it makes sense – although the term harks back to the days when the only people who drank wine were those who could afford to keep it in the freezing-cold vaulted cellars beneath their 30-room townhouses.

Chilling
Drink your white wine as cold as you like. Cheap and characterless wines can be improved immeasurably if they are cold enough – the anaesthetising effect of the temperature removes all sense of taste.

Corked wine
Wine trade surveys reveal that far too many bottles are in no fit state to be sold. The villain is very often cited as the cork. Cut from the bark of cork-oak trees, these natural stoppers have done sterling service for 200 years, but now face a crisis of confidence among wine producers. A diseased or damaged cork can make the wine taste stale because air has penetrated, or musty-mushroomy due to a chemical reaction. These faults in wine, known as corked or corky, should be

immediately obvious, even in the humblest bottle, so you should return the bottle to the supplier and demand a refund. A warning here. Bad corks tend to come in batches. It might be wise not to accept another bottle of the same wine, but to choose something else.

Today, more and more wine producers are opting to close their bottles with polymer bungs. Some are designed to resemble the 'real thing' while others come in a rather disorienting range of colours – including black. There seems to be no evidence that these synthetic products do any harm to the wine, but it might not be sensible to 'lay down' bottles closed with polymer. The effects of years of contact with these materials are yet to be scientifically assessed.

Corkscrews

The best kind of corkscrew is the 'waiter's friend' type. It looks like a pen-knife, unfolding a 'worm' (the screw) and a lever device which, after the worm has been driven into the cork (try to centre it), rests on the lip of the bottle and enables you to withdraw the cork with minimal effort. These devices are cheaper and longer-lasting than any of the more elaborate types, and are equally effective at withdrawing the new polymer bungs.

Decanting

Some wines need decanting because they are unfiltered and leave a gritty deposit in the bottom of the bottle. Cheaper wines suffer less from this admirable quirk, but there are plenty of rugged, budget bottles that will benefit from being sploshed into a new container – which could just as easily be another, clean wine bottle (use a funnel) as an elegant decanter. 'Airing' red wines in this way has real benefits, as distinct from the pointless 'breathing' exercise described above.

Keeping it

How long can you keep an opened bottle of wine before it goes downhill? Not long. A recorked bottle with just a glassful out of it should stay fresh until the day after, but if there is a lot of air inside the bottle, the wine will oxidise, turning progressively stale and sour. Wine 'saving' devices which allow you to withdraw the air from the bottle via a punctured, self-sealing rubber stopper are variably effective, but don't expect these to keep a wine fresh for more than a couple of re-openings. A crafty method of keeping a half-finished bottle is to decant it, via a funnel, into a clean half bottle and recork.

Storing it

Supermarket labels always seem to advise that 'this wine should be consumed within one year of purchase'. I think this is a wheeze to persuade customers to drink it up quickly and come back for more. Many of the more robust red wines are likely to stay in good condition for much more than one year, and plenty will actually improve with age. On the other hand, it is a sensible axiom that inexpensive dry white wines are better the younger they are.

If you do intend to store wines for longer than a few weeks, do pay heed to the conventional wisdom that bottles are best stored at low, stable temperatures, preferably in the dark. Bottles closed with conventional corks should be laid on their side lest the corks dry out and shrink for lack of contact with the wine.

Glasses

Tricky one, this. Does it make any difference whether you drink your wine from a hand-blown crystal glass or Old Mother Riley's hobnail boot? Do experiment! Conventional wisdom suggests that the ideal glass is clear, uncut, long-stemmed and with a tulip-shaped bowl large enough to hold a generous quantity when filled only halfway up. The idea is that you can hold the glass by its stalk rather than by its bowl. This gives an uninterrupted view of the colour, and prevents you smearing the bowl with your sticky fingers. By filling the glass only half-way up, you give the wine a chance to 'bloom', showing off its wonderful perfume. You can then intrude your nose into the air space within the glass, without getting it wet, to savour the bouquet. All harmless fun, really – and quite difficult to perform if the glass is an undersized Paris goblet filled, like a pub measure, to the brim.

Washing up

If your wine glasses are of any value to you, don't put them in the dishwasher. Over time, they'll craze from the heat of the water. And they will not emerge in the glitteringly pristine condition suggested by the pictures on some detergent packets. For genuinely perfect glasses that will stay that way, wash them in hot soapy water, rinse with clean, hot water and dry immediately with a glass cloth kept exclusively for this purpose. Sounds like fanaticism, but if you take your wine seriously, you'll see there is sense in it.

Wine and food

Wine is made to be drunk with food, but some wines go better with particular dishes than others. It is no coincidence that Italian wines – characterised by soft, cherry fruit and a clean, mouthdrying finish – go so well with the sticky delights of pasta.

But it's personal taste rather than national associations that should determine the choice of wine with food. And if you prefer a black-hearted Argentinian Malbec to a lightweight Italian Barbera with your Bolognese, that's fine.

The conventions that have grown up around wine and food pairings do make some sense, just the same. I was thrilled to learn in the early days of my drinking career that sweet 'dessert' wines can go well with strong blue cheese. As I don't much like puddings, but love sweet wines, I was eager to test this match – and I'm here to tell you that it works very well indeed as the end-piece to a grand meal in which there is a choice between pud and cheese.

Red wine and cheese are supposed to be a natural match, but I'm not so sure. Reds can taste awfully tinny with soft cheeses such as Brie and Camembert, and even worse with goats' cheese. A really extravagant, yellow Australian Chardonnay will make a better match. Hard cheeses such as Cheddar and the wonderful Old Amsterdam (top-of-the-market Gouda) are better with reds.

And then there's the delicate issue of fish. Red wine is supposed to be a no-no. This might well be true of grilled and wholly unadorned white fish such as sole or a delicate dish of prawns, scallops or crab. But what about oven-roasted monkfish or a substantial winter-season fish pie? An edgy red will do very well indeed, and provide much comfort for those many among us who simply prefer to drink red wine with food, and white wine on its own.

It is very often the method by which dishes are prepared, rather than their core ingredients, that determines which wine will work best. To be didactic, I would always choose Beaujolais or summer-fruit-style reds such as those from Pinot Noir grapes to go with a simple roast chicken. But if the bird is cooked as *coq au vin* with a hefty, winey sauce, I would plump for a much more assertive red.

Some sauces, it is alleged, will ovewhelm all wines. Salsa and curry come to mind. I have carried out a number of experiments into this great issue of our time, in my capacity as consultant to a company which specialises in supplying wines to Asian restaurants. One discovery I have made is that forcefully fruity dry white wines with keen acidity can go very well indeed even with fairly incendiary dishes. Sauvignon Blanc with Madras? Give it a try!

I'm also convinced, however, that some red wines will also stand up very well to a bit of heat. The marvellously robust Argentinian reds that get such frequent mentions in this book are good partners to Mexican chilli-hot recipes and salsa dishes. The dry, tannic edge to these wines provides a good counterpoint to the inflammatory spices in the food.

Some foods are inimical to wine. What would you choose, for example, with a boiled egg? Or a Nutella sandwich? My answer is to avoid boiled eggs and Nutella sandwiches whenever possible.

ADNAMS

This famous Suffolk brewery's wine-merchant arm is one of the most esteemed in Britain. Operating an efficient nationwide mail-order service, they offer an impressive range of 'fine' wines, but they have a large selection, too, of what they like to refer to as 'quaffing' wines. And these cheaper bottles, be assured, are just as carefully chosen as the grand wines they offer at the other end of the considerable Adnams price scale.

Buying 'everyday' wines from a merchant like Adnams makes sense, simply because this firm can be entirely trusted to sell only good wines. In 20 years of buying wine from them (mostly at the humblest end of the range) I have never had a bad bottle. But I have had plenty of excitingly delicious ones. Adnams have two wine shops, one in their seaside home of Southwold, the other in Norwich, and do all their other retail business by mail order. Their twice-yearly price list, lavishly illustrated and entertainingly written, is a proverbial Aladdin's cave. They do excellent sampling cases, including 'Superstar' selections of a dozen different wines at under £5 a bottle. Minimum order is a dozen bottles, and delivery is free to UK mainland addresses, reliably within a week of ordering.

If you have a growing interest in wine, Adnams' list will serve as a marvellous primer. It guides you persuasively up the price scale through a fabulous selection of wines you'll never see in supermarkets, all the way to the world's greatest and most expensive names. Prepare to be seriously tempted!

Adnams Wine Merchants, The Crown, High Street, Southwold, Suffolk IP18 6DP.

Tel: 01502 727220. Fax: 01502 727223. E-mail: wines@adnams.co.uk

Ⓥ Special for value Ⓠ Special for quality and interest Ⓥ̄Q̄ Special for value, quality and interest

RED LIGHT-MIDDLEWEIGHTS

£4.50		Figaro 1998	*Endearingly plausible, soft-fruit vin de pays de l'Hérault (despite silly name)*
£4.75	Ⓠ	Terrasses de Guilheim 1998	*Lean and minty but vigorously ripe Rhône vin de pays from a grand producer*
£4.85		Lar de Barros Tempranillo 1997	*Unoaked Spanish (Estramadura region) red with easy charm*
£4.95		Baso Navarra Garnacha 1998	*Lightweight Spanish red that gets a firm grip on your tastebuds; good with paella*
£4.95	Ⓠ	Roxan Montepulciano d'Abruzzo 1997	*Delicious, deep-purple, brambly Italian with sweet centre and dry finish*
£4.95	Ⓠ	Saumur Les Nivières 1997	*Edgy, well-defined red from a famed Loire Valley appellation; distinctive*

RED HEAVYWEIGHTS

£4.25	ⓋⓆ	Piornos Vinho Regional Beiras 1997	*Portuguese mid-heavyweight with sunny, blackberry fruit and tight finish; 13% alcohol*

SPLASHING OUT ON RED WINES

£5.50	ⓋⓆ	Salice Salentino, Vereto, 1995	*Beautifully coloured, mature and ripely rounded, earthy red from Apulia, Italy*
£8.50	Ⓠ	Château Thivin, Côte de Brouilly, 1998	*Like to discover real Beaujolais? This chewy-juicy fruit-rush is a brilliant introduction*

PINK

£4.95		Domaine Grange Rouge Syrah Rosé 1998	*Emphatic, brightly summer-fruited pink vin de pays d'Oc; good food matcher*

WHITE LIGHT-MIDDLEWEIGHTS

£3.50 **V** Jean des Vignes Vin de Pays du Gers
Adnams' 'house' dry white is fresh and bright, a high standard at this price

£3.95 Cépage Colombard, Côtes de Gascogne, 1998
Modern-method, crisp, dry vin de pays from good Plaimont co-op in Gascony

£4.95 **Q** Basa Rueda 1998
Northern Spanish, perfumed with toffee and ripe green fruit; exciting, lingering flavour

£4.95 Ozidoc Sauvignon 1998
Oddly named but recognisably squeaky-clean, green-fruit Sauvignon vin de pays

WHITE HEAVYWEIGHTS

£4.95 Domaine de Montmarin Marsanne, Vin de Pays Côte de Thongue, 1998
'Flavours of lime blossom, pears and greengage,' says Adnams; fair comment

SPLASHING OUT ON WHITE WINES

£6.95 **VQ** Château le Chec 1997
Divine dry white Graves (Bordeaux) with pure elegance from a great vineyard

£7.35 **Q** Mitchell Watervale Riesling 1998
Pure, piercing nose and mouthwatering limey fruit to this Australian classic

£7.95 **VQ** Forrest Sauvignon Blanc 1998
Top New Zealand pure-fruit quality; as good as Cloudy Bay and very much cheaper

PERSONAL NOTES:

....
....
....
....
....
....
....

ASDA

Asda has come a long way from the homely Associated Dairies origins which gave it its curious name. Now ranking as one of the Big Four supermarket chains, it is, of course, part of the biggest retailer in the world – Wal-mart. Once confined to northern England, Asda branches had already spread to most of the country (excepting Northern Ireland) before the takeover by the Americans announced in the summer of 1999, and it seems very likely this chain will grow to be perhaps the biggest in the country. There are plenty of rumours about what changes will be made at Asda, and it is probable that next year's edition of *Best Wine Buys* will show a very different range of wines from this.

The appearance of a huge United States retailer in Britain has all our supermarket companies rattled. Prices charged on this side of the Atlantic have astonished American executives, who have every intention of trying to scoop a very much larger share of this market for their new British possession. In future, perhaps we can look forward – no doubt with mixed feelings – to Asda branches on the scale of Wal-marts in the USA. Some stock 300,000 different product lines – ten times what you expect to find in a British store.

A more immediate prospect is that the 'price wars' that occasionally break out between supermarkets look set to be stepped up. Given the very high grocery prices we pay in UK supermarkets compared to other countries – for wine as well as everything else – this can only be a welcome development.

Ⓥ Special for value Ⓠ Special for quality and interest ⓋⓆ Special for value, quality and interest

RED LIGHT-MIDDLEWEIGHTS

£3.29	Ⓥ	Asda Chilean Red 1998	*Sleek, juicy and easy drinking; cracking quality at the price*
£3.99		Duro Douro 1997	*Juicy and bright Portuguese red from the same valley, and the same grapes, as port*
£3.99		Argentinian Tempranillo 1998	*Ripe, edgy, dark-fruit style from dependable La Agricola estate; vegetarian/vegan*
£3.99	Ⓥ	Svischtov Special Reserve Cabernet Sauvignon 1993	*Light-bodied but poised and ripe, mature Cabernet from Bulgaria*

RED HEAVYWEIGHTS

£3.99	Ⓠ	Argentinian Pinot Noir 1997	*Heady and full (13.5%), creamy red; hardly typical of Pinot, but very good*
£3.99		Arius California Zinfandel 1998	*Raisiny-strawberry nose, viscous strawberry fruit; very gulpable and 13.5% alcohol*
£3.99		Montenegro Vranac 1996	*Balkan red with bitter-cherry nose and dark fruit; robust red with staying power*
£4.99	Ⓠ	Peter Lehmann Vine Vale Grenache 1997	*Monster (14.5%), briary-ripe red from Barossa Valley, Australia*

PERSONAL NOTES:

. . . .
. . . .
. . . .
. . . .
. . . .
. . . .
. . . .
. . . .
. . . .
. . . .
. . . .
. . . .

£2.99		Asda Argentinian White 1998	All-purpose, cheap dry white with freshness and balance; vegetarian/vegan
£3.29	(V)	Asda Australian Oak Aged White	Quaint courgette-pineapple nose and easy fruit from usually boring Colombard grapes
£3.99	(Q)	Devil's Rock Riesling 1997	A good German; bright, keen style, not at all sweet, appley fruit, citrus edge
£3.99		D'Istinto Catarratto Chardonnay 1998	Australian (Hardys) inspired Sicilian dry white with tang, length and a frisky finish
£3.99		Prieuré Saint André Sauvignon 1998	Easy, balanced dry white Bordeaux
£4.99	(VQ)	Denis Marchais Hand Picked Vouvray 1998	Loire classic; fresh, mineral, honey flavours search out every taste bud; 11.5% alcohol

WHITE HEAVYWEIGHTS

£3.49		Côtes du Rhône Aged in Oak 1998	Soft, full, dry white with fragrant honeyed melon fruit
£4.69		Penfolds Rawson's Retreat Bin 21 Semillon-Chardonnay 1998	Generous Australian, dry style but heaps of lush, peachy, limey, fresh fruit
£4.99 ½ bottle	(Q)	Cranswick Nine Pines Vineyard Marsanne 1998	Exotic, lushly fruity, sweet Australian; 'pudding' wine but good as an aperitif
£4.99		Lindemans Bin 65 Chardonnay 1998	Ubiquitous but dependably generous and flavour-packed Australian

FIZZ

| £4.99 | (V) | Asda Cava | Spanish fizz with big bubbles, greeny colour and fruit; appley; good, and cheap |

BALLS BROTHERS

Well known to Londoners for its 20 distinctive wine bars and restaurants, this memorably named firm – founded 150 years ago by publican Henry Balls and still run by his descendants today – is also a renowned wine merchant. They have no shops, but operate a highly efficient nationwide mail-order service based on a catalogue issued twice yearly.

Delivery is free in central London for orders of a dozen or more bottles and free elsewhere in England, Wales and the Scottish lowlands on orders of two dozen or more (or orders worth above £150). The delivery service is, in my experience, very efficient – orders arrive within five working days.

Balls Brothers Ltd, 313 Cambridge Heath Road, London E2 9LQ.
Tel: 0171 739 1642. Fax: 0171 729 0258. Web: http://www.ballsbrothers.co.uk

V Special for value **Q** Special for quality and interest **(VQ)** Special for value, quality and interest

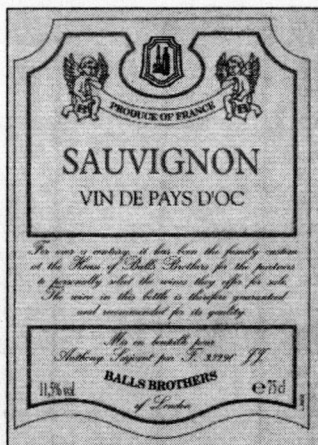

Balls Brothers' own-label Sauvignon Blanc has a keen acidity that will match it well with fish and chips – but go easy on the vinegar.

RED LIGHT-MIDDLEWEIGHTS

£4.50		Beaujolais AC, Depagneux, 1997	*Respectable example of the bouncy Beaujolais style; try it cool*
£4.50		Château Pech Céleyran, la Clape, 1997	*Languedoc red with sweet aromas and summery fruit*
£4.95	(V)	Balls Brothers Claret	*Leafy but ripe, Merlot-dominated Bordeaux by reliable producer Sichel; good value*

RED HEAVYWEIGHTS

£4.50		Copertino, Ruggieri 1996	*Rugged indeed, this lively-earthy mature red from deep-south Italy*
£4.50		Rosso del Salento, Santa Barbara, 1994	*Typically roasty, dark and clean-finishing southern Italian; ageing well*
£4.75	(V)	Domaine de Sainte Eulalie, AC Minervois, 1997	*Warm-fruit, deep-southern French with sunny appeal; consistent favourite*

SPLASHING OUT ON RED WINES

| £7.95 | (Q) | Mitchell Peppertree Vineyard Shiraz 1997 | *Spicy by name and nature, explosively fruity, intense, 14% alcohol from Clare Valley, Australia* |

WHITE LIGHT-MIDDLEWEIGHTS

£3.25	(V)	West Peak Sauvignon/Chenin Blanc 1997	*Serviceable and remarkably cheap South African fresh-soft glugger*
£3.75		Cépage Colombard, Côtes de Gascogne, 1998	*Modern-method, crisp, dry vin de pays from good Plaimont co-op in Gascony*
£3.95		Los Vilos Chardonnay 1998	*Brisk Chilean (Concha Y Toro) with grassy nose and a crunchy, appley fruit*
£4.05		Marqués de Cáceres Blanco 1997	*Modern-style white Rioja; brisk, fresh, unserious and definitely dry*

£4.25		Balls Brothers Sauvignon Blanc	*Proper briny, green-grassy style from this vin de pays d'Oc refresher*
£4.25	(VQ)	Domaine de Rieux 1998	*Brilliantly zingy, fresh, dry white from Gascony by cult wine-armagnac makers Grassa*
£4.25		Sunnycliff Colombard/ Chardonnay 1998	*Cheerfully scented, soft-dry style with refreshing easiness from Victoria, Australia*
£4.95	(Q)	Balls Brothers Alsace Pinot Blanc, Domaine Blanck, 1997	*Fresh and crisp, but exotic and tangy too; absorbing Alsatian of real interest*
£4.95		Sunnycliff Sauvignon Blanc 1998	*Nice bloom of gooseberry and keen, faintly tropical fruit; from Victoria, Australia*

WHITE HEAVYWEIGHTS

£4.50	(V)	Domaines Virginie Marsanne 1998	*Perfumed vin de pays d'Oc; dry with attractive canned fruit nuances; works well*

SPLASHING OUT ON WHITE WINES

£6.75	(VQ)	Mitchell Watervale Riesling 1998	*Pure, piercing nose, and mouth-watering limey fruit to this Australian classic*
£7.75	(Q)	Leasingham Domaine Chardonnay 1996	*Muscular Australian; peachy-tropical character and yellow richness*
£7.95	(Q)	Chablis Domaine des Manants 1997	*Limpid gold-green colour, pebbly-limey nose, zesty, complex fruit; pure Chablis*

BERKELEY WINES

see WINE CELLAR

BIBENDUM

This smart London merchant, based in Regent's Park, is a major importer of wines from Europe and the New World. As the agent for numerous producers right around the globe, Bibendum supplies many brands to supermarkets and restaurants. But you can buy these wines direct from them by mail order, at keen prices. They publish a glossy, comprehensive list at least once a year, and operate an efficient home-delivery service, which is free for orders of one case or more in mainland England and Wales. In Scotland and offshore England and Wales, the charge is £10 for one to four cases, and free for five cases or more. Deliveries are next day in London and within three working days in mainland England and Wales.

Bibendum Wine Ltd, 113 Regent's Park Road, London NW1 8UR.

Tel: 0171 916 7706. Fax: 0171 916 7705. Web: www.bibendum-wine.co.uk

Ⓥ Special for value Ⓠ Special for quality and interest ⓋⓆ Special for value, quality and interest

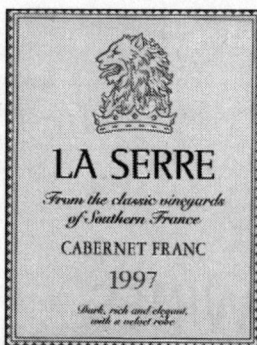

La Serre Cabernet Franc's ripe fruit flavours will suit red meats with sauces, such as Steak Béarnaise.

RED LIGHT–MIDDLEWEIGHTS

£3.50		Casa Rosso, Tenuta Casalbaio, 1998	*Light (11.5%), but excitingly pure and ripe red from Puglia, Italy*
£3.50	(VQ)	Libertad Malbec/Sangiovese 1998	*Argentinian with silkiness and lush cherry-blackberry fruit; exciting, and cheap*
£3.50		Uvello Rosso 1997	*Simple, berry-fruit Italian; slurp it cool as a party red or with sticky pasta dishes*
£3.75	(Q)	La Piazza Rosso, Tunuta Casalbaio, 1997	*Vibrant, characterful Sicilian glugger with intensity and heart*
£3.95	(VQ)	Colori Sangiovese di Puglia 1997	*Another Casalbaio wine from Italy's deep south; fine cherry and heather style*
£3.95	(Q)	Mision Tinto 1998	*Amazing, soft and slurpable red from Bodegas Santo Tomas – in Mexico*
£4.50	(Q)	Ceps du Sud Old Vine Mourvèdre 1997	*Redcurrants and burnt rubber! – fascinatingly delicious vin de pays d'Oc*
£4.50		Marquis du Gard 1997	*Bibendum's 'house' claret is a good tooth-coater with firm fruit; not bad at the price*
£4.75		St Chinian Mas de Berre 1996	*From France's deep south, a structured, mature red, jam-packed with flavours*

RED HEAVYWEIGHTS

£3.95	(V)	Colori Primitivo di Puglia 1997	*Creamy-rich and splendidly black-cherry-style red by Casalbaio in Italy's deep south*
£3.95		1885 La Rural Malbec 1998	*From Argentina, a dusky, muscular mouthfiller*
£3.95		La Serre Cabernet Franc 1997	*Opulent southern French with plum and raspberry aroma*
£4.95	(Q)	Allora Primitivo di Puglia 1998	*Another Casalbaio wine, an enriched, intensified edition of the Colori above*

| £4.95 | (VQ) | Deakin Estate Merlot 1998 | *Velvety, powerful (14% alcohol) Australian with lush, black-cherry sweetness* |
| £4.95 | (VQ) | Deakin Estate Shiraz 1998 | *Lipsmacking, juicy, red-fruit Australian; remarkable value; 14.5% alcohol; good with curry!* |

WHITE LIGHT-MIDDLEWEIGHTS

£3.75	La Piazza Trebbiano di Sicilia 1998	*Light and fresh young wine with a typically sunny Sicilian disposition*
£3.95	Muscadet des Ducs, Chereau Carré, 1998	*Lively and assertively acidic example of this famed Loire wine; genuine quality*
£4.95	Muscadet sur Lie, Pierre Guindon, 1998	*Fine, fresh wine with an extra dimension of stony-crisp fruit; fish partner*
£4.95	Valdivieso Chardonnay 1998	*Plenty of flavour in this vigorous, fresh Chilean made with Bibendum participation*

WHITE HEAVYWEIGHTS

£4.95	Ceps du Sud Viognier 1998	*Heady aroma of spring flowers and dried apricots; full, soft and smoky style*
£4.95	Springfield Estate Barrel Fermented Chenin Blanc 1998	*From Robertson, South Africa, a creamy, exotic, dry refresher*
£4.95	Valdivieso Sauvignon Blanc 1998	*Heavyweight by Sauvignon standards; hefty gooseberry fruit from Chile*

SPLASHING OUT ON WHITE WINES

| £5.95 | Alfred Barrel Fermented Chardonnay 1998 | *Remarkable, luminously fruity, rich wine from Australia's Deakin estate* |
| £8.95 | Catena Agrelo Vineyard Chardonnay 1996 | *Yes, expensive, but a great classic wine you'll never forget; from Argentina* |

£4.75   Grand Imperial   *Very cheap French tank-method fizz, but it's fresh and quaffable; good mixer*

PERSONAL NOTES:

....
....
....
....
....
....
....
....
....
....
....
....
....
....

*Valdivieso Chardonnay has the vitality to stand up to assertive fish dishes,
including* moules marinière.

BOOTHS

An independent chain of 24 supermarkets in northern England, Booths stands out as an admirable anachronism in the era of the huge multi-national retail corporation. The company was founded by Edwin Henry Booth, who opened his first shop at Blackpool in 1847. He was then aged 19 and had survived a childhood marked by unspeakable cruelties.

In the best traditions of Victorian self-made men, Booth sustained his integrity and purposefulness in spite of ejection from his home, aged only ten, by a murderously drunken stepfather, exploitation by a series of Dickensian employers and years of deprivation. His own account of his life, published under the title *Shadow and Sheen* in 1897 and reissued by the company to mark its 150th anniversary, is a touching and enlightening piece of social history.

Booths is still owned and managed by Edwin's descendants, and is a comfort to behold as a beacon of independent retailing at the turn of the millennium. They have a cracking range of wines, and prices are keen. As with other supermarkets there are regular promotions on all sorts of wines and if you buy six bottles of any regular mix of wines at once, you get a 5 per cent discount. As definitely not with other supermarkets, if your wine purchase is worth more than a total of £150, you get a whopping 15 per cent discount. The stores are mostly in the founder's home county of Lancashire, but there are branches, too, at Ilkley in Yorkshire and at Kendal, Ulverston and Windermere in Cumbria.

V Special for value   **Q** Special for quality and interest   **VQ** Special for value, quality and interest

RED LIGHT-MIDDLEWEIGHTS

£2.89	V	Alta Mesa Estremadura 1997	*Light but typically spicy-minty Portuguese at a very low price; drink cool*
£3.19		Merlot del Veneto, Zonin	*Very light, cherry-style young red from north-west Italy – and very cheap*
£3.25	V	Château Laval, Costières de Nîmes, 1997	*Earthy, gripping middleweight from a great southern French AC; real value*
£3.69		Cahors, Côtes d'Olt, 1997	*Cheap for this dark-fruit red from the famous AC's biggest co-operative*
£3.89	Q	Libertad Malbec/Sangiovese 1998	*Argentinian with silkiness and lush cherry-blackberry fruit; exciting, and cheap*
£4.39		Biovinum Côtes du Rhône, Louis Mousset, 1997	*Organic wine with more interest than usual from Côtes du Rhône; lively fruit*
£4.49		Kumala Cinsault-Pinotage 1998	*South African middleweight; earthy-sweet, slightly raisiny fruit is an acquired taste*
£4.55		Valdivieso Merlot 1998	*From a reliable Chilean producer, well-structured, sweet-earthy stuff*

RED HEAVYWEIGHTS

£3.89	VQ	Portada Red 1997	*Big, sweet, minty Portuguese of tremendous quality; look for the hint of honey!*
£4.29	VQ	Belafonte Baga 1997	*Pitch-dark Portuguese with a big burst of berry fruit; weighty, good with spicy food*
£4.29	Q	Château Veyran, St Chinian, 1996	*A personal favourite; sweet, Mediterranean-herb warmth from the Languedoc*
£4.59		La Rural Malbec 1998	*From Argentina, a dusky, muscular mouth-filler*

£4.69	Q	Cono Sur Cabernet Sauvignon 1997	*Intense, leafy-blackcurranty wine from Chile*
£4.69	V	Cono Sur Pinot Noir 1997	*Nice raspberry nose to this earthy Chilean burgundy-style red*
£4.99	VQ	Dão Dom Ferraz 1997	*Dark colour, dark fruit to this rich and powerful (13% alcohol) Portuguese*
£4.99	VQ	Salice Salentino, Vallone, 1996	*Fine garnet colour, dark, mildly scorched fruit; terrific interest and value from Italy*

SPLASHING OUT ON RED WINES

| £5.99 | VQ | Redwood Trail Pinot Noir 1997 | *Elegantly delicious Californian red convincingly akin to burgundy at twice this price* |
| £8.49 | Q | The Twenty-eight Road Mourvèdre, d'Arenberg, 1996 | *Gorgeous monster velvet McLaren Vale red; a chance to try Australia's best* |

WHITE LIGHT-MIDDLEWEIGHTS

£3.19	V	Honoré de Berticot Semillon, Côtes de Duras 1998	*Alluring tropical bouquet to a fresh, dry southern French white; good fruit at the price*
£3.29		Retsina Kourtaki	*Cheap but fresh and typically pine-resiny reminder of all those taverna nights*
£3.49		Vinho Verde, Caves Aliança	*Medium, faintly spritzy Portuguese; less trendy than once, but still fun if cold*
£3.79		Chapel Hill Oaked Chardonnay 1997	*Pretty good lightweight with appley-toasty undercurrent, from Lake Balaton, Hungary*
£3.99	V	Niersteiner Spiegelberg Riesling Spätlese, Rudolf Miller, 1997	*Complex orchardy fruit in this ripe and racy late-picked Rhine wine*
£3.99	Q	Penfolds Rawson's Retreat Bin 202 Riesling 1998	*Australian Riesling, nothing like German; lively, gooseberry-like zing*

| £4.39 | Marqués de Cáceres Blanco 1997 | *Modern-style white Rioja; brisk, fresh, unserious and definitely dry* |
| £4.99 | Mâcon-Lugny Eugène Blanc 1997 | *Real white burgundy under a fiver; rare and a good, healthy Chardonnay* |

WHITE HEAVYWEIGHTS

£4.49	Alsace Pinot Blanc, Turckheim, 1998	*Fresh and flowery style from one of Alsace's top co-operative producers*
£4.49	Penfolds Rawson's Retreat Bin 21 Semillon-Chardonnay 1998	*Generous Australian, dry style but heaps of lush, peachy, limey, fresh fruit*
£4.59 ●	Hardys Stamp Series Riesling/ Gewürztraminer 1998	*Delicious spicy-peachy Australian that is inexplicably thinly distributed*
£4.99 Ⓥ	Cordillera Estate Chardonnay Reserva 1997	*Lavishly coloured wine from Casablanca, Chile, with purity and elegance*
£4.99 ●	Deakin Estate Chardonnay 1998	*Impressive yellow and lush style from Victoria, Australia*
£4.99	James Herrick Chardonnay 1998	*Much advertised, but nevertheless satisfyingly big-fruited, balanced vin de pays d'Oc*
£4.99	Lindemans Bin 65 Chardonnay 1998	*Ubiquitous but dependably generous and flavour-packed Australian*

SPLASHING OUT ON WHITE WINES

£5.99 ●	Viña Esmeralda 1997	*Grapey-lychee-smoky, brilliantly exotic but refreshingly dry white by Torres in Spain*
£7.99 ●	Dashwood Sauvignon Blanc 1998	*Sea-fresh Kiwi, loaded with gooseberry, green-grass fruit; even better than the '97*
£8.49 ●	Jackson Estate Sauvignon Blanc 1998	*Brilliant, briny, classic Sauvignon from New Zealand; as good as it gets*

£3.99 (V)	Cavalier Brut Blanc de Blancs	*French-made Euro-fizz is dry but not without relishable flavour; very cheap*
£4.99	Palau Brut	*This is a cava, from Penedès in Spain; delicious, crisp refresher*
£4.99	Prosecco Brut, Zonin	*Commercial but easy-to-drink north-east Italian spumante; drink it seriously chilled*

PERSONAL NOTES:

....
....
....
....
....
....
....
....
....
....
....
....
....
....
....
....

BOOZE BUSTER

see WINE CELLAR

BOTTOMS UP

see FIRST QUENCH

BUDGENS

Budgens is a chain founded in 1872, now with 115 outlets in south-central and south-east England. A number of the stores are incorporated into petrol stations and the company plans to continue in this direction. This review of Budgen wines is rather brief, as I did not embark on it until shortly before going to press, and an offer by the company's representative to send me some of the Budgen own-brand wines for tasting was not fulfilled.

Comparing prices, as one inevitably does in compiling a guide like this, there is something noticeable about Budgens. For widely available branded wines, their prices are in some cases the lowest anywhere.

V Special for value **Q** Special for quality and interest **VQ** Special for value, quality and interest

LA Cetto's Petite Sirah is from Mexico, but is a subtle, soft red. Drink with rice or pasta dishes rather than chilli!

RED LIGHT-MIDDLEWEIGHTS

£3.99	ⓞ	JP Chenet Cabernet Syrah 1997	*Pleasant vin de pays middleweight in an endearing olde-worlde wonky bottle shape*
£3.99		Viña Albali Reserva 1993	*Lightish colour but plenty of interest; mature, sweet, vanilla-toned; Valdepeñas, Spain*
£4.32		Kumala Cinsault-Pinotage 1998	*South African middleweight; earthy-sweet, slightly raisiny fruit is an acquired taste*
£4.49		Waimanu Dry Red	*From New Zealand; has a likeable, eucalyptus, sappy-silk style to the red fruit*

RED HEAVYWEIGHTS

£3.99		Faugères Jeanjean 1997	*Robust Languedoc (French deep-south) with a bit of grip*
£4.39	ⓋⓆ	Dão Dom Ferraz 1997	*Dark colour, dark fruit to this rich and powerful (13% alcohol) Portuguese*
£4.39	Ⓥ	LA Cetto Petite Sirah 1996	*Good price for this well-coloured, oaky, Spanish-style red from Mexico*
£4.99	Ⓥ	Hardys Stamp Series Shiraz-Cabernet Sauvignon 1998	*Solid South Australian with firm, lipsmacking cassis fruit; above-average value*

PERSONAL NOTES:

....
....
....
....
....
....
....
....
....
....

WHITE LIGHT-MIDDLEWEIGHTS

| £2.99 | Ⓥ | Retsina Kourtaki | *Cheap but fresh and typically pine-resiny reminder of all those taverna nights* |

WHITE HEAVYWEIGHTS

£4.49	Ⓥ	Hardys Stamp Series Chardonnay-Semillon 1998	*Good price for this emphatically flavoured Australian*
£4.69		Penfolds Rawson's Retreat Bin 21 Semillon-Chardonnay 1998	*Generous Australian, dry style but heaps of lush, peachy, limey, fresh fruit*
£4.99		James Herrick Chardonnay 1998	*Much advertised but nevertheless satisfyingly big-fruited, balanced vin de pays d'Oc*
£4.99		Lindemans Bin 65 Chardonnay 1998	*Ubiquitous but dependably generous and flavour-packed Australian*

PERSONAL NOTES:

....
....
....
....
....
....
....
....
....
....
....
....
....
....

CELLAR 5

see WINE CELLAR

CO-OP

Great changes are afoot in the confusingly structured Co-operative movement. Until 1999, there were two dominantly large Co-op organisations among the many societies around the country: the Co-operative Retail Society (CRS) and the Co-operative Wholesale Society (CWS). Each had its own buying and distribution department, serving two quite separate retail networks. But now, in an administrative merger which has been said to have cost thousands of Co-op workers their jobs, these two societies are unifying their purchasing activities. This will mean that the thousand-or-so groceries supplied by the combined organisation will sell a common range of wines, all selected by the Co-operative Wholesale Society.

The wines mentioned here are distributed by this centralised Co-op department, which supplies more than 90 per cent of the shops around the country, including the Pioneer superstores and Stop & Shop branches. Obviously, the range carried in any individual outlet depends on the size of the branch in question.

Note that the Co-op regularly offers substantial discounts on their wines on a month-to-month basis, so it's always worth perusing the shelves for special offers when you visit your local branch.

Ⓥ Special for value ⬤ Special for quality and interest ⓋⓆ Special for value, quality and interest

£3.49	Co-op Argentine Malbec-Bonarda 1998	*Clean finish to this keenly priced, rather Italian-style, young red*
£3.49	Winter Hill Red 1998	*A pale and interesting vin de pays d'Oc; squishy summer fruit*
£3.99	Graffigna Shiraz-Cabernet 1998	*Well-coloured middleweight Argentinian; firm fruit and a bit of spice*
£3.99	Viña Albali Reserva 1993	*Lightish colour but plenty of interest; mature, sweet, vanilla-toned Valdepeñas (Spain)*

RED HEAVYWEIGHTS

£3.99	Co-op Chestnut Gully Monastrell-Merlot 1998	*Hearty Spanish (Yecla, SE Spain) dark-fruit red by an Australian winemaker*
£3.99	Co-op Vin de Pays Syrah-Malbec	*Solid young briary red from the South of France; vegetarian*
£4.99 ⓠ	Trulli Primitivo 1997	*Fierce dark fruit in this vibrant red from Salento, Italy*
£4.99 ⓠ	Balbi Vineyard Cabernet Sauvignon 1997	*Quality from Mendoza, Argentina; firm, structured, silky Cabernet; classic*
£4.99	Terra Mater Zinfandel/Shiraz 1998	*Hearty (14% alcohol) Chilean blend; unchallenging but relishable, soft-fruit style*

SPLASHING OUT ON RED WINES

| £5.99 ⓥⓠ | Redwood Trail Pinot Noir 1997 | *Elegantly delicious Californian red; convincingly akin to burgundy at twice this price* |

PERSONAL NOTES:

....
....
....
....
....

WHITE LIGHT–MIDDLEWEIGHTS

£3.29		Co-op Hungarian Chardonnay 1997	*Light but not 'flabby' (wine-speak for watery) bargain dry white with pineapple whiff*
£3.79		Gyöngyös Estate Chardonnay 1998	*Recognisable Chardonnay with hints of cream and apple; Hungarian*
£3.99		Butterfly Ridge Sauvignon-Chenin Blanc 1998	*Soft but dry lightweight by Angove's in Australia; hint of kiwi fruit*
£3.99		Co-op Vin de Pays Chardonnay-Chenin Blanc	*Little hint of sweet pear in the fruit of this southern-French softie; vegetarian*
£3.99	Q	Devil's Rock Riesling 1997	*A good German; bright, keen style, not at all sweet, appley fruit, citrus edge*

WHITE HEAVYWEIGHTS

£3.99	Q	Rio de Plata Torrontes 1998	*Exotic, grapey but nicely lime-edged wine from Argentina*
£4.29		Long Mountain Semillon-Chardonnay 1998	*Big-selling South African with peachy power; by Jacob's Creek creator Robin Day*
£4.49	V	Hardys Stamp Series Chardonnay-Semillon 1998	*Good price for this emphatically flavoured Australian*
£4.69		Penfolds Rawson's Retreat Bin 21 Semillon-Chardonnay 1998	*Generous Australian, dry style but heaps of lush, peachy, limey, fresh fruit*

FIZZ

£4.49	Co-op Asti Spumante	*Sweet but fresh Italian sparkler to drink thoroughly chilled*
£4.99	Co-op Cava Brut	*Plenty of colour and persistent fizz; Spanish, with more heart than most at this price*

Trulli Primitivo's emphatic acidity and firm fruit will make it a good choice with pizza – pepperoni included.

PERSONAL NOTES:

....
....
....
....
....
....
....
....
....
....
....
....
....
....

DRINKS CABIN

see FIRST QUENCH

FIRKIN OFF LICENCE

see FIRST QUENCH

FIRST QUENCH

Not a name you recognise? It is, you might be surprised to know, the biggest wine-merchant chain in the country. The network is formed by Thresher Wine Shops, the Whitbread-owned chain of off-licences, incorporating Wine Rack shops and Bottoms Up 'wine superstores' amalgamated with Allied Domecq's Victoria Wine. First Quench is not a name you'll see above any of the shops (not yet, anyway), which continue to trade, mostly, under their original names.

The effect of this mega-merger has been to produce a truly huge nationwide network of shops. At the time the deal was announced, back in 1998, the two companies had nearly 3,000 outlets between them. Inevitably, there have been and will be more branch closures where shops are too immediately adjacent, but the 'estate' will still be by far the largest in the country.

First Quench are rationalising the wine range, too. All the shops will sell wines from one central list. You'll find pretty much the same choice not just in Bottoms Up, Thresher, Wine Rack and Victoria Wine, but in all the other outlets included in the new chain – Drinks Cabin, Firkin Off Licence, Haddows, Hutton's and Martha's Vineyard.

Not unnaturally, the company has been accused of perpetuating the process of conglomerating Britain's once diverse collection of high street wine shops into one monolithic whole. But First Quench managers tell me they have no intention of continuing to diminish the diversity of names either over their shops or on their shelves. The official word is: 'We believe we are improving quality, not reducing choice.'

Only time will tell. Meanwhile, the wines in this section should be available in your local Bottoms Up, Martha's Vineyard, Thresher, Victoria Wine or Wine Rack branch. These various subdivisions of the First Quench empire have varying special terms on purchases of a more than a few bottles of wine, sometimes connected to membership (free) of 'clubs'. Do always ask about discounts – you should expect at least 10 per cent off if you buy a dozen bottles at a time. And do take advantage of the money-off coupons on pages 155–60.

Ⓥ Special for value Ⓠ Special for quality and interest ⓋⓆ Special for value, quality and interest

RED LIGHT-MIDDLEWEIGHTS

£3.49		Ed's Red	*Everyday Spanish with better fruit and balance than the gimmicky packaging suggests*
£3.79	(VQ)	Libertad Malbec/Sangiovese 1998	*Argentinian with lush cherry-blackberry fruit; exciting, and cheap for this quality*
£3.99		Don Darias	*'Oaky' Spanish red, not quite the bargain it once was*
£3.99	**V**	Terra Boa, Tras-Os-Montes 1998	*Portuguese middleweight with lively, grippy, mature fruit; very good value*
£3.99		Fitou Special Reserve 1998	*Ripe and meaty stuff from the reliable Mont Tauch co-op in France's Midi*

RED HEAVYWEIGHTS

£4.49		La Palma Cabernet-Merlot 1997	*Rich colour, plenty of cassis and plum fruit, hint of tobacco; from Rapel, Chile*
£4.49		Primitivo Trulli 1997	*Earthy, faintly scorched, dark, intense red from Salentino in Italy's sun-baked deep south*
£4.99		Cono Sur Pinot Noir 1997	*Nice raspberry nose to this earthy Chilean burgundy-style red*
£4.99	(VQ)	Dão Dom Ferraz 1997	*Dark colour, dark fruit to this rich and powerful (13% alcohol) Portuguese*
£4.99		Domaine Boyar Premium Oak Barrel Aged Merlot 1997	*Black-cherry nose on this sweetly ripe and densely fruited Bulgarian*
£4.99		Norton Malbec 1997	*Grippy, ripe and soft red from renowned estate in Mendoza, Argentina*
£4.99	**Q**	Oak Village Cabernet Sauvignon 1996	*Muscular South African with pure blackcurrant nose; dense and mature*
£4.99	(VQ)	Salice Salentino, Vallone 1996	*Fine garnet colour, dark, mildly scorched fruit; terrific interest and value from Italy*

£4.99		Valdivieso Merlot 1998	*From a reliable Chilean producer, well-structured, sweet-earthy stuff*

SPLASHING OUT ON RED WINES

£5.99	(VQ)	Redwood Trail Pinot Noir 1997	*Elegantly delicious Californian red, convincingly akin to burgundy at twice this price*

PINK

£3.99		Cool Ridge Pinot Noir Rosé 1998	*Lurid magenta colour; likeable sweet-but-crisp style; Bottoms Up and Wine Rack only*

WHITE LIGHT-MIDDLEWEIGHTS

£2.99		Casa Rural White	*Spanish vino de mesa; clean, fresh style with some fruit – and cheap*
£3.49	V	The Unpronounceable Grape 1998	*It's Hungarian and called the Cserzegi Fuszeres; fresh and crisp but softly fruity*
£3.69		Las Colinas Chilean White	*Dry style, plenty of fruit here, though*
£3.99	Q	Penfolds Rawson's Retreat Bin 202 Riesling 1998	*Australian Riesling, nothing like German; lively, gooseberry-like zing*
£3.99		Pinot Grigio, Fiordaliso 1998	*Superior example of this trendy Italian dry white; smoky, thirst-slaking style*
£4.99	Q	La Baume Sauvignon Blanc 1998	*The nose positively sings on this lush, fruit-laden vin de pays d'Oc*
£4.99		Orvieto Secco Antinori 1997	*Florally scented and herbaceous dry wine from famed hill town in Umbria, central Italy*
£4.99	Q	Orvieto Abboccato Antinori 1997	*Soft and gently unctuous (rather than sweet) version of the secco wine above; delightful*

WHITE HEAVYWEIGHTS

£4.49	🄌	Norton Torrontes 1998	*Grapey, exotic Argentinian with a very soft style; drinks well with spicy foods*
£4.69	🄌	Hilltop Gewürztraminer 'Slow Fermented' 1997	*Whiff of lychee and lush, pungent flavours from Hungary*
£4.69		Penfolds Rawson's Retreat Bin 21 Semillon-Chardonnay 1998	*Generous Australian, dry style but heaps of lush, peachy, limey, fresh fruit*
£4.99		James Herrick Chardonnay 1998	*Much advertised but nevertheless satisfyingly big-fruited, balanced vin de pays d'Oc*
£4.99		Lindemans Bin 65 Chardonnay 1998	*Ubiquitous but dependably generous and flavour-packed Australian*

SPLASHING OUT ON WHITE WINES

£5.99	🄌	Viña Esmeralda 1997	*Grapey-lychee-smoky, brilliantly exotic but refreshing dry white by Torres in Spain*
£6.99	🄌	Dashwood Sauvignon Blanc 1997	*Fresh as sea air and loaded with gooseberry, green-grass fruit; New Zealand*

PERSONAL NOTES:

....
....
....
....
....
....
....
....
....
....
....
....
....
....

FULLER'S

The famed London brewery has a chain of 62 wine shops in the capital and home counties. These are pleasant places in which to browse, but you need to take your time – there are hundreds of wines crammed into what can sometimes seem a rather small space. Strong points for bargain hunters are Italy, Spain and South America. Fuller's claims to have 'nearly three times the national average in terms of Chilean wine mix', and for Argentina 'four times the national percentage of mix'. I take this to mean they have more Chilean and Argentinian wines than other shops. As with other high street chains, there is a constant stream of in-store promotions, usually based on seven bottles for the price of six, buy two bottles and save a pound or two, and so on. Always ask about discounts for purchases of a dozen or more bottles.

V Special for value **Q** Special for quality and interest **VQ** Special for value, quality and interest

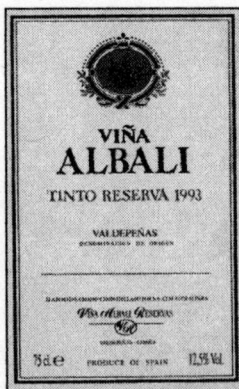

From the 1993 vintage, Viña Albali is deliciously mature, and just the red to drink with simple meat dishes – or even bangers and mash.

RED LIGHT-MIDDLEWEIGHTS

£2.99	(V)	Rustica Rosso 1998	*This rustic red is from Sicily; a likeable warmly earthy red without faults – and cheap*
£3.59		Winter Hill 1998	*A pale and interesting vin de pays d'Oc; squishy summer fruit*
£3.75		J&F Lurton Tempranillo/ Bonarda 1997	*Spanish-Italian grape blend gives lively fruit; from Mendoza, Argentina*
£3.75	(VQ)	Libertad Malbec/Sangiovese 1997	*Argentinian with lush cherry-blackberry fruit; exciting, and cheap*
£3.99	(V)	Espiral Tempranillo/Cabernet 1997	*From Somontano, Spain, an impressive blackcurranty slurper; pity about the naff label*
£4.49	(Q)	Alianza Merlot 1997	*Baked fruit flavours, Marmite character, voluptuous soft fruit in this Argentinian*
£4.99	(Q)	Barbera D'Asti, Alasia, 1995	*Zingy, brambly, juicy red from a leading producer in Piedmont, Italy*
£4.99		Cono Sur Pinot Noir 1997	*Nice raspberry nose to this earthy Chilean burgundy-style red*

RED HEAVYWEIGHTS

£4.19	(Q)	Santa Ines Cabernet/Merlot 1997	*From Chile, a brightly ripe Bordeaux-style with a satisfying finish*
£4.29	(V)	Concha Y Toro Merlot 1998	*From Chile's biggest producer, a consistently delicious and well-priced Merlot*
£4.85		Vine Vale Grenache, Peter Lehmann, 1997	*Monster (14.5% alcohol) briary-ripe red from Barossa Valley, Australia*
£4.99	(VQ)	Alamos Ridge Cabernet Sauvignon 1995	*Lavish mature 'cigar box' Argentinian by a top producer; £1 cheaper than elsewhere*

£4.99		Norton Cabernet Sauvignon 1997	*Satisfying, ripe blackcurrant style from a dependable estate in Mendoza, Argentina*
£4.99		Casillero del Diablo Merlot 1997	*Dark and spicy red from Concha Y Toro of Chile*
£4.99	Q	Cono Sur Cabernet Sauvignon 1997	*Intense, leafy-blackcurranty wine from Chile*
£4.99	V	Hardys Stamp Series Shiraz-Cabernet Sauvignon 1998	*Solid South Australian with firm, lipsmacking cassis fruit; above-average*
£4.99		Isla Negra Cabernet Sauvignon 1996	*Forcefully fruity Chilean; this will stand up to salsa*
£4.99	Q	Viña Albali Reserva 1993	*Mature (look at the vintage), creamy-oaky Rioja-style from Valdepeñas, Spain*
£4.99		Wild Pig Reserve Shiraz 1996	*Vin de pays d'Oc; softly, solidly delicious rather than a hairy brute*

SPLASHING OUT ON RED WINES

| £5.99 | VQ | Redwood Trail Pinot Noir 1997 | *Elegantly delicious Californian red convincingly akin to burgundy at twice this price* |
| £7.99 | VQ | Quinta do Crasto Reserva 1995 | *Great super-ripe but elegantly poised red from Portugal's Douro; an experience* |

PERSONAL NOTES:

....
....
....
....
....
....
....
....
....
....
....
....

WHITE LIGHT-MIDDLEWEIGHTS

£3.99 ⓞ Concha Y Toro Gewürztraminer 1998 — *Soft, off-dry grapey white from Chile; notes of grapefruit and melon; fun*

£3.99 La Vega Verdejo/Sauvignon 1998 — *Dry whites from Rueda in Spain are consistently zesty and delicious; try this one*

£4.59 João Pires Dry Moscato 1998 — *Exotic Portuguese dry white with floral nose and grapey-melon fruit; aperitif wine*

£4.99 (VQ) Big Frank's Viognier 1998 — *Great name and a pleasing, soft, figgy-tropical fruit; has Viognier character*

£4.99 Muscadet sur Lie, Domaine de la Fruitière, 1998 — *Superior Muscadet with the usual bracing, slaking style, but some flavour, too*

£4.99 Valdivieso Chardonnay 1998 — *Generous flavour has been extracted into this healthy, fresh Chilean*

WHITE HEAVYWEIGHTS

£4.69 Penfolds Rawson's Retreat Bin 21 Semillon-Chardonnay 1998 — *Generous Australian, dry style but heaps of lush, peachy, limey, fresh fruit*

£4.99 Alsace Pinot Blanc 1998 — *Floral style and freshness from Alsace's dependable Turckheim co-op*

£4.99 Lindemans Bin 65 Chardonnay 1998 — *Ubiquitous but dependably generous and flavour-packed Australian*

£4.99 Nottage Hill Riesling 1997 — *Australian Riesling with appley-mineral emphasis; not at all like German Riesling*

£7.99 (VQ)   Catena Agrelo Vineyard   *Cheaper than elsewhere for this*
              Chardonnay 1997   *great classic Argentinian white*

PERSONAL NOTES:

....
....
....
....
....
....
....
....
....
....
....
....
....
....
....
....
....
....
....
....
....

HADDOWS

see FIRST QUENCH

HUTTON'S

see FIRST QUENCH

KWIK SAVE

This nationwide chain of 830 stores was taken over by Somerfield in 1998. Some Kwik Saves are being converted to the Somerfield 'banner' but many continue to trade under the original name as the four-year project to absorb the chain into Somerfield is completed. Wine-sourcing has been centralised under the control of Somerfield, so you can expect to find many of the same wines in both chains. Own-brand wines in Kwik Save are, for the most part, already under the Somerfield label. See the Somerfield section starting on page 103.

At Kwik Save, own-label wines have now been largely replaced by those of the Somerfield range.

LONDIS

KWIK SAVE

ondis wines might not enjoy a high publicity profile, but the range is available in more shops than any other chain – around 1,850 outlets. The shops are everywhere in England and Wales, on city street corners, in sleepy villages, in the market squares of country towns. The name Londis may not appear above the shop door – many go under their proprietors' own banners – but you soon know from the countless branded products on sale within that you're in a Londis outlet.

There is a big range of wines, both Londis own-labels and well-known brands, for the shops to choose from. All the wines reviewed here are on the Londis list, and there should be a good few of them in your local shop. Prices given here, which look pretty competitive, are recommended retail, but may vary.

Take advantage of the discount voucher on page 157.

V Special for value **Q** Special for quality and interest **VQ** Special for value, quality and interest

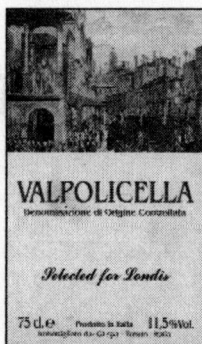

Londis Valpolicella is good of its kind, and is an inexpensive accompaniment to pasta dishes.

RED LIGHT-MIDDLEWEIGHTS

£3.29	**V**	Londis Côtes du Roussillon	*Good, firm, spicy, ripe fruit in this middleweight from France's deep south; a bargain*
£3.39		Londis Minervois	*Light, but right-tasting wine from this well-known AC of France's Midi*
£3.49		Vallade Merlot del Veneto 1997	*Very lightweight but has a keen, edgy acidity that suits sticky pasta dishes*
£3.59		Vallade Sangiovese di Puglia 1997	*Identifiably Italy's Sangiovese grape (of Chianti fame), pleasingly abrasive; needs food*
£3.79	**V**	Londis Valpolicella	*This works well; plenty of colour and nutskin-dry finishing fruit; good pasta partner*
£4.29		Londis Beaujolais	*Bricky colour and mature style; rather a good quaffer, though untypical of Beaujolais*
£4.49		Waimanu Dry Red	*From New Zealand; has a likeable eucalyptus, sappy-silk style to the red fruit*
£4.99	(VQ)	Rio de Plata Cabernet Sauvignon 1995	*Pure, silky, structured Cabernet from Argentina*

RED HEAVYWEIGHTS

£3.49	**V**	Londis Fitou	*Well-chosen southern French AC with warm fruit and clean finish*
£4.99		Long Mountain Cabernet Sauvignon 1997	*Pretty decent South African; ripe but not overheated; stands up to assertive food*
£4.99		Viña Porta Cabernet Sauvignon 1997	*Slightly tough Chilean, well structured and up-front; try with bangers and mash!*

SPLASHING OUT ON RED WINES

| £5.99 | (VQ) | Redwood Trail Pinot Noir 1997 | *Elegantly delicious Californian red convincingly akin to burgundy at twice this price* |
| £7.49 | (Q) | Wolf Blass Yellow Label Cabernet Sauvignon 1997 | *Big (13% alcohol) Australian – gaudy livery and explosive, mega-ripe fruit; great stuff* |

WHITE LIGHT-MIDDLEWEIGHTS

£1.99		Londis White Lambrusco	*Unnerving kinship with lemonade, but inoffensive and only 4% alcohol*
£2.79 litre		Londis Hock	*German tafelwein; clean, grapey, okay; cheapest full-strength (9% alcohol) wine in this book*
£3.29		Piesporter Michelsberg, Zellerbard Kellerei, 1998	*Bright appley style to this basic Moselle; drink chilled; only 8.5% alcohol*
£4.99		Nobilo White Cloud 1998	*Old-fashioned 'medium' New Zealand white, but with likeable, fresh balance*

WHITE HEAVYWEIGHTS

£3.99	(V)	Rio de Plata Torrontes 1997	*Exotically perfumed dry Argentinian with appealing grapey style; fun*
£4.49		Penfolds Rawson's Retreat Bin 21 Semillon-Chardonnay 1998	*Generous Australian, dry style but heaps of lush, peachy, limey, fresh fruit*
£4.89		Hardys Stamp Series Chardonnay-Semillon 1998	*Emphatically flavoured Australian with more character than most sub-£5 Oz brands*
£4.99		Lindemans Bin 65 Chardonnay 1998	*Ubiquitous but dependably generous and flavour-packed Australian*

MAJESTIC

The original 'wine warehouse' where the wares are stacked in canyons of cartons has been one of the great success stories of recent times. From an original single branch bought in 1980 by investment banker Giles Clarke (who sold out long ago and is now one of Britain's richest men), Majestic has grown into a major force in the market with 82 outlets nationwide – and plans to reach 150 within the next five years.

It seems extraordinary that no other retailer has copied the original idea. But then Majestic isn't a retailer, it's a wholesaler. This is why you have to buy at least 12 bottles of wine at a time. And in times of less liberal licensing hours, it was how Majestic branches could stay open all day, seven days a week.

They've lost that edge now, but have earned loyalty from a large army of regular customers who like to buy their wine by filling an oversized supermarket-style trolley steered between the canyons. Prices are pretty good, and the company continually discounts large numbers of wines. They publish regular price lists, and it's sensible to study the current one before making a visit, because it's all too easy in these crowded places to overlook some of the best buys. Majestic's particular strengths include German wines. They have an uncanny knack of finding mature, high-quality Rieslings, often priced under £5. Look out, too, for the cheapest mature clarets to be found anywhere. Majestic seek these out from unlikely sources such as Scandinavian liquor monopolies selling off surplus stock.

If you don't have a Majestic within reasonable driving distance, you can order wine for home delivery. Minimum order is a dozen bottles and delivery is free on the UK mainland. Look up your nearest Majestic in the phone book and ring to order. If there isn't a branch in your directory – there are only three branches in Scotland (in Glasgow and Edinburgh), one in Wales (Cardiff) and none in the north-east – telephone head office on 01727 847935.

Ⓥ Special for value Ⓠ Special for quality and interest Ⓥ̄Q̄ Special for value, quality and interest

RED LIGHT-MIDDLEWEIGHTS

£2.99	Basilicata Rosso 1997	*Convincingly cheap midweight with respectable fruit from Italy's 'toe'*
£3.49	Pedras do Monte 1997	*Budget Portuguese with concentration and juicy fruit*
£3.99 Ⓥ	Côtes du Rhône Caves des Papes 1996	*Plump, mature fruit in this oak-aged bargain in an expensive-looking bottle*
£4.49	Correas Syrah Sangiovese 1997	*Generous ripe fruit and an austere but manageable backtaste in this Argentinian*
£4.49	Periquita, JM da Fonseca 1996	*Easy-drinking Portuguese; mature fruit with interesting hints of spice*
£4.79	Valpolicella Classico, Santepietre, 1998	*Proper cherry style to this refreshingly fruity Italian classic; finishes well*
£4.99	Beaujolais AC, Georges Duboeuf, 1998	*Identifiable bouncy Beaujolais and a pretty, floral label; not great, but fun*

RED HEAVYWEIGHTS

£4.49 Ⓞ	Pupilla Cabernet Sauvignon, Luis Felipe Edwards, 1998	*From a good Chilean producer; pure fruit, bags of character; like it*
£4.99	Château Guiot, Costières de Nîmes, 1998	*Soft, juicy heart and grippy texture; from one of the Languedoc's best appellations*
£4.99	Col di Sasso 1997	*Pleasingly abrasive but richly fruited Tuscan by Yankee mega-winery Banfi*
£4.99 Ⓞ	Meia Pipa 1993	*Voluptuous, super-ripe Portuguese smoothie with mint, spice and velvet*
£4.99	Ochoa Tempranillo Garnacha 1998	*Muscular Spanish from Navarra (Rioja's junior neighbour) with assertive flavours*

| £4.99 | (VQ) | Prosperity Red | Californian, mainly Cabernet, with chocolatey black-fruit richness; 13.5% alcohol |

SPLASHING OUT ON RED WINES

| £7.49 | ❶ | Wolf Blass Yellow Label Cabernet Sauvignon 1997 | Big (13% alcohol) Australian – gaudy livery and explosive, mega-ripe fruit; great stuff |
| £8.99 | ❶ | Chianti Rufina Nippozzano Riserva 1996 | Dense, chewy, classic black-cherry Chianti for that very special Italian meal |

PINK

| £4.99 | | Santa Rita Cabernet Sauvignon Rosé 1998 | A rare thing – a really delicious rosé; salmon colour, fresh cassis fruit; from Chile |

WHITE LIGHT-MIDDLEWEIGHTS

£3.79		Chardonnay-Pinot Grigio delle Venezie, Pasqua, 1998	Very pale Italian from an odd grape blend but fine floral nose and cheery fleshiness
£3.99		Chenin Remy Pannier 1998	Loire Valley dry white with a hint of tinned peaches and brisk limey finish
£3.99		Domaine de Fontanelles Sauvignon 1998	Convincing, gooseberry-fresh vin de pays d'Oc
£3.99		Sauvignon Lot 279 1998	Pleasantly brisk, dry white Bordeaux with clean finish
£4.49		João Pires Dry Moscato 1997	Exotic Portuguese with flowery, grapey appeal
£4.99	(VQ)	Ockfener Bockstein Riesling Kabinett 1992	Gorgeous appley Moselle; whiff of orange sponge cake; only 8% alcohol
£4.99	❶	Bianco di Custoza, Cavalchina, 1998	Brisk Italian in the Soave style; hint of blanched almonds, limey finish

£4.99		Bourgogne Chardonnay Meilleurs Climats 1997	*White burgundy for under a fiver is a rarity; this gives an idea of the style*
£4.99		Frascati Superiore, Fontana Candida, 1998	*The wine of Rome from its most famous producer; not bad value for the quality*
£4.99		Ironstone Vineyards 'Obsession' 1998	*Californian oddity, very grapey style; good in spite of groovy presentation*
£4.99	Q	Oberemmeler Rosenberg Riesling Kabinett 1993	*Whiff of petrol on this smoky, limey Moselle classic*
£4.99	VQ	Wiltinger Hölle Riesling Kabinett, Jacobus, 1990	*Marvellous mature Moselle with classic 'petrolly' style of old Riesling*

WHITE HEAVYWEIGHTS

£4.69		Penfolds Rawson's Retreat Bin 21 Semillon-Chardonnay 1998	*Generous Australian, dry style but heaps of lush, peachy, limey, fresh fruit*
£4.99	VQ	Domaine de Raissac Viognier 1998	*Raspberries, figs and apricots feature in this superb, rich vin de pays d'Oc*
£4.99	Q	Domaine de Raissac CVM Barrique 1998 Vin de Pays d'Oc 1998	*Chardonnay/Viognier/Muscat grapes aged in oak equals an exotic treat; 13% alcohol*
£4.99		Lindemans Bin 65 Chardonnay 1998	*Ubiquitous but dependably generous and flavour-packed Australian*
£4.99	Q	Prosperity White 1998	*Green-gold, tinned-pineapple nose, toffee-apple fruit; Firestone Vineyard, California*

SPLASHING OUT ON WHITE WINES

| £6.99 | ⓋⓆ | Armand Riesling Kabinett, Von Buhl, 1997 | *Spectacularly fruity, racy, German Riesling of unforgettable power; in a black bottle* |
| £8.99 | Ⓠ | Jackson Estate Sauvignon Blanc 1998 | *Brilliant, briny, classic Sauvignon from New Zealand; as good as it gets* |

SHERRY

| £4.99 | Ⓥ | Hidalgo Fino | *Excellent, dry and tangy pale sherry from a great producer; bargain price; drink cold* |
| £4.99 | | Hidalgo Amontillado | *Attractive, off-dry, nutty style to this authentic, balanced sherry; drink it cool* |

PERSONAL NOTES:

....
....
....
....
....
....
....
....
....
....
....
....
....
....
....
....
....
....
....
....
....
....
....

MARKS & SPENCER

The first 'supermarket' to introduce own-label wines 30 years ago under the St Michael brand has now lost its way among the leaders. The range is comparatively tiny – only about 150 still wines and very few 'fine' wines. But among the standard fare – much of it very disappointing – there are a number of decent wines worth trying. Nine out of ten bottles cost under £5, so the choice on a budget is arguably as good as it is in stores with wider choices.

M&S offer a discount if you buy 12 bottles. The slogan is, 'Buy a dozen and get the twelfth bottle free'. What this really means is that if you buy 12 bottles costing a total of, say £48, you get a twelfth (8.5 per cent) off – and thus pay £44. It's a better deal than you'd get in any of the big supermarket chains, where standard discounts are limited to 5 per cent on six or more bottles.

Finally, do take note that if you are buying a dozen wines, you won't be able to pay by credit card and if your purchase is over £50, don't bank on a cheque guarantee card either. If you haven't got an M&S card, you'll need cash or a debit card such as Switch.

V Special for value **Q** Special for quality and interest **VQ** Special for value, quality and interest

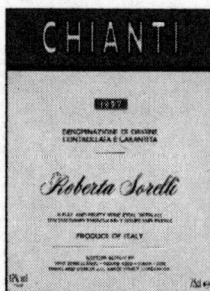

Marks & Spencer Chianti is unusually well priced – and will drink very well with chicken dishes and pasta.

RED LIGHT-MIDDLEWEIGHTS

£2.99	House Red Wine 1998	*Southern French* vin de table *with respectable fruit but price is principal attraction*
£3.99 litre	Italian Red 1998	*Quaffable vegetarian/vegan vino da tavola; litre price equates to £3 for a 75cl bottle*
£3.99	Reggiano Rosso 1998	*Soft but well-defined, from Emilia-Romagna region, Italy; easy drinking*
£4.49	Montepulciano d'Abruzzo Ponte d'Oro 1997	*Screwtop; delightful cherry-fruit Italian*
£4.99	Chianti Roberta Sorelli 1997	*Cheap (for Chianti) with typical keen fruit and a nice dry edge; good buy*

RED HEAVYWEIGHTS

| £4.49 | Gold Label Cabernet Sauvignon 1997 | *Vegetarian vin de pays d'Oc with bracing style; good with spicy food* |

PERSONAL NOTES:

....
....
....
....
....
....
....
....
....
....
....
....
....
....
....
....

WHITE LIGHT-MIDDLEWEIGHTS

£2.99	V	Vin de Pays du Gers 1998	Crisp style with emphatic fruit and citrussy edge – and cheap; vegetarian/vegan
£3.49		Bianco di Puglia 1998	Easy, fresh, dry white from southern Italy
£4.99		Mandeville Viognier 1998	Grassy style, soft but dry vin de pays from southern France

WHITE HEAVYWEIGHTS

| £4.99 | Sierra Los Andes Chardonnay 1998 | From Carmen winery, Chile; well made with lush fruit |

PERSONAL NOTES:

....
....
....
....
....
....
....
....
....
....
....
....
....
....
....
....
....
....

MARTHA'S VINEYARD

see FIRST QUENCH

ODDBINS

They are the trendiest wine shops of them all, still managed by wild-eyed enthusiasts and decorated with blackboards advertising crazy price reductions and obscure new wines from outlandish places. So it's always sobering to contemplate that Oddbins is owned by one of the world's biggest multi-national drinks companies, the Canadian-based Seagram corporation.

It's a truly national chain, with 200 shops in England and Wales, 40 in Scotland and two in Ireland. There is even a branch in France, in the huge Cité Europe shopping centre in Calais, where you can buy the famous range at usefully reduced prices. The nationwide mail-order service, Oddbins Delivers, offers most of the wines available in the shops along with a good number that aren't.

Tel: 0870 601 0015.

(V) Special for value (Q) Special for quality and interest (VQ) Special for value, quality and interest

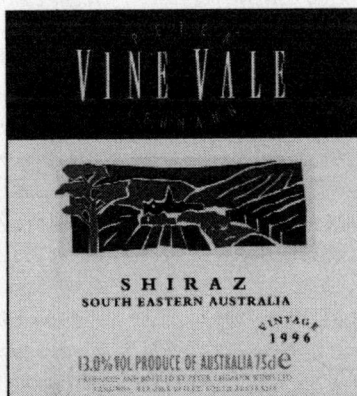

Peter Lehmann Vine Vale Shiraz, from the 1998 vintage, will make a very good match with roast lamb or shepherd's pie.

RED MIDDLEWEIGHTS

£3.49	(V)	Oddbins Red 1997	*Very respectable own-label vin de pays d'Oc; good mouthful for the money*
£3.69		Wild Pig 1998	*Name is the main attraction of this vin de pays d'Oc by gimmick-maestro Gabriel Meffre*
£3.99		Barton & Guestier Merlot 1997	*Decent middleweight vin de pays d'Oc; B&G, like Oddbins, is a Seagram company*
£3.99		Carta Vieja Cabernet Sauvignon 1997	*Dependable Chilean budget wine; middleweight and decent purity of fruit*
£3.99	(V)	Carta Vieja Merlot 1998	*Rather a supple and poised wine at this price; from Chile*
£4.69		Château Lartigue 1996	*From Bordeaux's satellite Côtes de Castillon AC, a crisply ripe, good-value claret*
£4.49		Kumala Cinsault-Pinotage 1998	*South African middleweight; earthy-sweet, slightly raisiny fruit is an acquired taste*

RED HEAVYWEIGHTS

£3.99		Métairie du Bois Syrah 1998	*Nicely concentrated and balanced, spicy vin de pays d'Oc*
£4.19		Dão Grão Vasco Tinto 1995	*Robust and old-fashioned mature Portuguese with sweet highlights*
£4.49		Primitivo Trulli 1997	*Earthy, faintly scorched, dark, intense red from Salentino in Italy's sun-baked deep south*
£4.49		Ptomaine des Blagueurs Grenache 1996	*Nicely rasping, dark Rhône red by quirky Californian winemaker Randall Grahm; fun*
£4.49	(Q)	Vila Regia Douro 1995	*Rich, glyceriny, minty, mature red from Portugal's Douro – where port comes from*
£4.69	(V)	Vine Vale Grenache, Peter Lehmann, 1998	*Monster Australian with 14.5% alcohol but very slurpable; red meat or spicy food partner*

£4.99		Casillero del Diablo Merlot 1997	*Dark and spicy red from Concha Y Toro of Chile*
£4.99		Cent'are Sicilia Rosso 1997	*Roasted, earthy style to this creamily oaked Sicilian; improves with airing*
£4.99		Cono Sur Pinot Noir 1998	*Hint of raspberry in this slightly austere Chilean; earthy*
£4.99	(VQ)	Deakin Estate Shiraz 1998	*Lipsmacking, juicy, red-fruit Australian; remarkable value; 14.5% alcohol; good with curry!*
£4.99		Isla Negra Cabernet Sauvignon 1996	*Forcefully fruity Chilean, this will stand up to salsa*
£4.99		Norton Malbec 1997	*Grippy, ripe and soft red from renowned estate in Mendoza, Argentina*
£4.99		Norton Merlot 1996	*Big, black-cherry softie with grip and a reassuring, long finish; Argentina; 13.5% alcohol*
£4.99	(V)	Sandeman Claret 1997	*Dependably decent, soft Bordeaux under the name of the famous port-sherry house*
£4.99		Viña Porta Cabernet Sauvignon 1998	*From Chile, a chewy but appealing Cabernet; vigorous cassis fruit*
£4.99		Vine Vale Shiraz, Peter Lehmann, 1998	*Big, jammy Australian with proverbial up-front fruit; heady stuff*

SPLASHING OUT ON RED WINES

£7.99	(Q)	Norton Privada 1996	*Top Argentine oaked red from a grand mélange of Bordeaux grapes; rich and smooth*
£7.99	(Q)	Chianti Classico, Castello di Volpaia, 1997	*Lovely cherry-and-almonds style to this great classic wine from a famed producer*

PINK

£4.99		Santa Rita Cabernet Sauvignon Rosé 1998	*A rare thing – a really delicious rosé; salmon colour, fresh cassis fruit; from Chile*

WHITE LIGHT-MIDDLEWEIGHTS

£2.49 (V)	Bereich Bernkastel	*Inoffensive and exceedingly cheap German plonk for the undemanding drinker*
£3.49	Nagyrede Dry Muscat 1997	*Cheery grapey whiff on this soft-fruited but clean white from Hungary*
£3.59	Retsina Kourtaki	*Cheap but fresh and typically pine-resiny reminder of all those taverna nights*
£3.79	Chapel Hill Oaked Chardonnay 1997	*Pretty good lightweight with appley-toasty undercurrent, from Lake Balaton, Hungary*
£3.99	Carta Vieja Sauvignon Blanc 1998	*Recognisable green, sea-breeze, Sauvignon style at a fair price; from Chile*
£3.99	Deidesheimer Hofstück Kabinett, St Ursula, 1997	*Decent, dry and mildly racy hock; aperitif wine*
£4.25	Vermentino di Sardegna, Sella & Mosca, 1996	*Endearing, flowery perfume to this dry but squishily fruity, white from Sardinia*
£4.99	Casablanca Sauvignon Blanc 1998	*Stacks of fruit, slightly green, very crisp; from a grand producer in Chile*

WHITE HEAVYWEIGHTS

£4.69	Penfolds Rawson's Retreat Bin 21 Semillon-Chardonnay 1998	*Generous Australian, dry style but heaps of lush, peachy, limey, fresh fruit*
£4.99	Histonium Chardonnay 1998	*Fun classical label on this nicely oaked Italian; complex nose, shades of toffee apples*
£4.99 (O)	Norton Torrontes 1998	*Grapey, exotic Argentinian with a very soft style; drinks well with spicy foods*

| £4.99 | | Pinot Blanc Cuvée Réserve 1998 | *From excellent Alsace co-op at Turckheim, slakingly delicious, exotic dry white* |

SPLASHING OUT ON WHITE WINES

| £7.99 | (VQ) | Fetzer Viognier 1997 | *Apricots, figs and almonds are suggested by this gorgeous Californian luxury white* |

PERSONAL NOTES:

. . . .
. . . .
. . . .
. . . .
. . . .
. . . .
. . . .
. . . .
. . . .
. . . .
. . . .
. . . .
. . . .
. . . .

PIONEER STORES

see CO-OP

RIGHT CHOICE

see WINE CELLAR

SAFEWAY

When I am asked which supermarket I believe has the best range of wines, I tend to hesitate. It's a toss-up between Waitrose and Safeway. But when it comes to wines under a fiver, there's no doubt about it. Safeway has a much larger range and makes a good job of putting a very wide selection of them even into its smaller branches. Many of these wines are 'own brands' and a good number are exclusive to Safeway.

What's more, Safeway continually offers useful price reductions on dozens of different lines, and initiated the 5 per cent discount on any six bottles that is now the norm for all the biggest supermarket chains.

So, Safeway is hereby pronounced the best supermarket of the lot. And as the number of enthusiastically recommended wines appearing here indicates, this chain clearly has the biggest and best range under a fiver.

(V) Special for value (Q) Special for quality and interest (VQ) Special for value, quality and interest

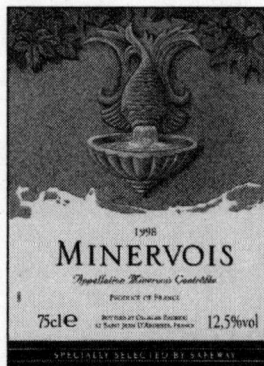

Safeway's own Minervois drinks wonderfully with traditional starchy dishes from southern France, such as cassoulet.

£2.79	Don Coyote Red 1997	*Blissfully named Spanish plonk; quite well balanced; plenty of bite; a 3-litre wine box costs £10.79, equivalent to £2.70 a bottle*
£3.19 (VQ)	Safeway Corbières 1998	*Scary colour; lovely bouncy nose and easy, friendly, ripe light-middleweight fruit*
£3.19 (VQ)	Safeway Minervois 1998	*Dark and sweet and cheerfully tannic young southern French red; brilliant value*
£3.29 (V)	Winter Hill 1998	*A pale and interesting vin de pays de l'Aude; squishy summer fruit giving real pleasure*
£3.49 (VQ)	Caballo de Plata Bonarda/Barbera 1998	*Lovely firm fruit to this easy-to-drink Argentinian from Italian grape varieties*
£3.49	Desert Vines Carignan 1997	*From Morocco; alarming Ribena colour but with a nice spike of pepperiness; 13% alcohol*
£3.49	Falua 1998	*From Portugal; dense colour, farmyardy nose and light in weight, but likeable*
£3.49	Mendoza Dry Red Wine 1997	*Clean and balanced Argentinian*
£3.49 (VQ)	Young Vatted Cabernet Sauvignon 1998	*Lovely burst of fruit in the mouth from this blackcurranty Bulgarian*
£3.49 (V)	Young Vatted Tempranillo 1997	*Squashed-raspberry nose and matching eager fruit in this Spanish bargain*
£3.79	Don Darias	*Pioneering oaky Spanish plonk now rather overtaken by New World rivals*
£3.99	Carta Vieja Merlot 1998	*Lively in the Beaujolais style, just a little woody; a cheery young Chilean*
£4.49 (Q)	Castillo de Sierra Rioja	*Nutty-sweet and identifiably Rioja (Spain) in style; easy middle-weight, but 13% alcohol*

£3.69		Apostolos Alentejo	*Big, plummy, agricultural number from Portugal; just the thing with a robust stew*
£3.99	Ⓠ	Domaine des Bruyères 1998	*Southern French with dense colour, impressive weight and easy, clean summer fruit*
£3.99	Ⓥ	Mendoza Merlot 1997	*Ripe, round Argentinian with bouncy, black-cherry fruit and 13.5% alcohol*
£4.29	ⓋⓆ	Safeway Mendoza Oak Aged Tempranillo 1998	*Ripe and ready red from reliable La Agricola in Argentina; vegetarian/vegan*
£4.49	Ⓠ	Belafonte Baga 1997	*Pitch-dark Portuguese with a big burst of berry fruit; weighty, good with spicy food*
£4.49		Malbec LA Cetto 1997	*Strong (13.5% alcohol), earthy-spicy Mexican; improves with airing*
£4.49	Ⓥ	Merlot Selection 24 1998	*Vin de pays d'Oc with plenty of tannin masking the warm, spicy fruit; rather grand*
£4.99	Ⓠ	Bourgueil Les Chevaliers 1997	*Tough but deeply interesting red from a once-fashionable Loire appellation; real quality*
£4.99	Ⓥ	Hardys Stamp Series Shiraz-Cabernet Sauvignon 1998	*Solid South Australian with firm, lipsmacking cassis fruit; above-average value*
£4.99		Mimosa Sangiovese di Maremma 1997	*Generous (13% alcohol) Italian from the Chianti grape; fine, dry finish; perfect with pasta*
£4.99		Terra Mater Merlot 1998	*From the Canepa winery in Chile; sweet briary nose and a nice 'cut'*
£4.99		Terra Mater Zinfandel/Shiraz 1998	*Hearty (14% alcohol) Chilean blend; unchallenging but relishable soft-fruit style*
£4.99		La Source Merlot/Syrah 1998	*Good vin de pays with a youthful leafiness; sweet middle fruit and a clean edge*

| £4.99 | 🅞 | Stellenbosch Cabernet Sauvignon 1998 | *Rather muscular South African; structured wine destined to improve with age* |

SPLASHING OUT ON RED WINES

£5.49		Domaine des Lauriers, Faugères, 1996	*A magnificent mouthful of summery fruit with a delectable bite of tannin on the edge*
£5.99	Ⓥ	Azbuka Merlot 1996	*Fine colour and elegant Bordeaux style to this outstanding Bulgarian wine*
£5.99		Fairview Pinotage 1998	*South African; big (13.5% alcohol), strawberryish red with intensity and heart; delicious*
£5.99		Isla Negra Merlot 1998	*Ripe black-cherry style with richness and dimension; special wine from Chile*

PERSONAL NOTES:

....
....
....
....
....
....
....
....
....
....
....
....
....
....
....
....
....
....
....

WHITE LIGHT–MIDDLEWEIGHTS

£2.99 (V) La Loustère, Vin de Pays du Gers, 1998
Tangy and keen, jolly, fresh and well made; rather a triumph, especially at the price

£2.99 (VQ) Woodcutter's White 1998
Crisp but seductively soft and very cheap dry white from Hungary; a bargain

£3.49 Hilltop Bianca 1998
Soft organic wine from the mysterious Bianca grape, grown in Neszmely, Hungary

£3.69 Safeway Bulgarian Oaked Chardonnay 1997
Lots of colour, nose with a hint of dried figs and sweet melon; coconut-vanilla note

£3.99 Riverview Chardonnay 1998
Hungarian; appley nose, peachy fruit and crisp acidity

£3.99 Safeway Matra Mountain Sauvignon Blanc 1998
Hungarian; gooseberry fruit – a bit tart, but zingily fresh; Sauvignon all the way

£3.99 Verdicchio dei Castelli di Jesi Classico, Moncaro, 1998
Pale and shy on the nose but nice minerally-nutty style; hint of spice at the finish; Italian

£4.79 (Q) Wild Trout 1998
Vin de pays d'Oc with breezy freshness and a crisply refreshing style; good with fish!

£4.99 (VQ) Marc Xero Chardonnay 1998
Minerally fruit, creamy depths in this screwtop-bottled bargain from Salento, Italy

WHITE HEAVYWEIGHTS

£3.99 Riverview Gewürztraminer 1999
Hungarian; lychee aroma; good weight, spice and smokiness

£4.49 (V) 35 Sur Sauvignon Blanc 1998
Beach-grass nose and plenty of dimension and interest to this Chilean classic

£4.69 Penfolds Rawson's Retreat Bin 21 Semillon-Chardonnay 1998
Generous Australian, dry style but heaps of lush, peachy, limey, fresh fruit

£4.99	(VO)	Cordillera Estate Chardonnay Reserva 1997	*Lavishly coloured wine from Casablanca, Chile, with purity and elegance; special*
£4.99	(Q)	Safeway Alsace Pinot Blanc 1998	*Herbaceous, subtly spicy mouthfiller from Alsace; good introduction to the style*
£4.99		Safeway Australian Oaked Chardonnay 1998	*Standard generous Oz 'shardy' at a standard price; clean finish*

SPLASHING OUT ON WHITE WINES

£5.99	(Q)	Basedow Semillon 1997	*Powerful Barossa, dry with luscious Semillon perfume, ripe fruit and crisp finish*
£6.99	(Q)	Vergelegen Chardonnay 1997	*Famed South African estate; elegant style, masses of fruit and 13.5% alcohol*

PERSONAL NOTES:

....
....
....
....
....
....
....
....
....
....
....
....
....
....
....
....
....
....
....
....

J SAINSBURY

Sainsbury's is, supposedly, in the doldrums. The loss of market leadership to Tesco has been followed by disappointing profits and the well-publicised cutting of thousands of management jobs. Will it all make the 400 supermarkets less attractive to shoppers?

When it comes to wine, don't give it a thought. Sainsbury's may no longer be Britain's biggest wine merchant, but the quality of their wines is still very consistent. One of the more remarkable features of their enormous range is the collection of bargain German wines. Their £2.79 Liebfraumilch, made by 'New World winemaking techniques' is a delightful wine by any standards, and a million miles from the dull, sugary style that has rightly pushed Liebfraumilch to the verge of extinction.

Sainsbury's is strong in the £4–5 sector, as the large selection of wines here demonstrates. As with other big chains, bear in mind that only the very largest branches carry more than a proportion of the total number of wines on the company's long list.

In common with the other big chains, Sainsbury's does frequent promotions on wines, and offers a 5 per cent discount on purchases of six or more bottles at a time.

V Special for value **Q** Special for quality and interest **VQ** Special for value, quality and interest

£2.99	(V)	Mendoza Country Red	*Cherry fruit but plenty of dimension and depth to this Italian-style Argentinian*
£3.29	(V)	Sainsbury's Bordeaux Rouge	*Leafy, young-fruit style to this pure and enjoyable claret; unusual quality and value*
£3.29	(V)	Sainsbury's Syrah Vin de Pays d'Oc	*Vividly fruity with softness and a pleasing silky texture; good at this price*
£3.49		Sainsbury's Merlot Corvina 1998	*Bouncy, juicy young Italian; sweet, but not too sweet*
£3.99	(Q)	Bright Brothers Merlot/Tannat 1998	*From Uruguay! Leafy, tobacco whiff; sweet, ripe black-cherry fruit*
£3.99		LPA Côtes de St Mont 1998	*Dense southern French with hedgerow nose and easy tannin*
£3.99		Sainsbury's South African Merlot	*Pale; likeable nutty-summer-fruit style; warm-weather red*
£3.99	(V)	Terra Boa, Tras-Os-Montes, 1997	*Portuguese middleweight with lively, grippy, mature fruit; very good value*
£4.49		Alteza 750 Tempranillo	*Sweetly ripe but quite mouth-puckering Spanish red; try with the barbecue*
£4.49		Alteza 775 Tempranillo Cabernet Sauvignon	*Strange colour, but genuinely pleasing raspberry style to this vigorous Spanish wine*
£4.99	(V)	Beaujolais Villages Les Roches Grillées 1998	*Rare item – recognisable Beaujolais at under £5; nice inky-juicy style*
£4.99		Château Le Noble 1996	*Bordeaux petit château with a ripe, sweet-berry style*
£4.99		Mimosa Barrel Aged Sangiovese 1997	*Chianti-style, pretty good, dry-finishing sweetish red from Tuscany, Italy*
£4.99	(VQ)	Rio de Plata Cabernet Sauvignon 1995	*Pure, silky, structured Cabernet from Argentina*

£3.99	Q	Bright Bros Atlantic Vines Baga 1997	*Brambly, dark-fruit and meaty (13% alcohol) red from Portugal; big wine*
£4.49		Alteza 600 Old Vines Garnacha	*Almost syrupy in weight (13% alcohol) and intensity; dark and spicy flavours from Spain*
£4.49		Cawarra Shiraz/Merlot/Ruby Cabernet 1998	*From Lindeman's (Australia); briary nose, sweet fruit, tannic bite*
£4.49	Q	Merlot delle Venezie, Connubio, 1998	*Bright, cherry-spicy Italian; pleasantly mouth-puckering pasta-matcher*
£4.99		Bankside Gallery Merlot 1997	*Sweet and approachable vin de pays d'Oc with concentration (13% alcohol) and length*
£4.99	V	Bankside Gallery Syrah 1997	*Purply young vin de pays d'Oc; spicy nose, sinewy fruit; tannic but not tough; good*
£4.99		Bright Brothers Vistalba Malbec	*Generous, balanced Argentinian red with plenty of weight and interest*
£4.99		Chasse du Pape 1997	*Big-brand Rhône, rather pale, but with intensity, warm spice and proper slurpability*
£4.99	VQ	Dama de Toro 1998	*From Toro, Spain; muscular stuff, but smooth and silky with obvious oak*
£4.99		Domaine Boyar Premium Oak Barrel Aged Merlot 1997	*Black-cherry nose on this sweetly ripe and densely fruity Bulgarian*
£4.99		James Herrick Cuvée Simone 1997	*Startlingly vivid colour and burstingly ripe fruit in this big-name vin de pays d'Oc*
£4.99	V	Hardys Stamp Series Shiraz-Cabernet Sauvignon 1998	*Solid South Australian with firm, lipsmacking cassis fruit. Above-average value*
£4.99		Mota del Cuervo Premium Tempranillo 1997	*Spanish with handsome colour and cool, lush, summer-fruit and a long finish*

£4.99		Premium Oak Bulgarian Merlot 1997	*Dense, New-World-style Bulgarian with generous black-cherry style*
£4.99	V	Terra Mater Zinfandel Syrah 1998	*Dark and dense (14% alcohol) Chilean with purity and insinuating fruit; easy drinking*
£4.99	Q	Trilogie, Maurel Vedau, 1996	*Vin de pays d'Oc, a little 'green' but structured with spice and grunt (13% alcohol)*

SPLASHING OUT ON RED WINES

| £5.99 | VQ | Alamos Ridge Cabernet Sauvignon 1996 | *Glorious, accessible wine from Catena, arguably Argentina's best producer; class act* |
| £9.99 | Q | Château La Cardonne, Médoc, 1995 | *Yes, double the budget, but a rare chance in a supermarket to try lovely, mature claret* |

PINK

| £2.99 | | Sainsbury's Romanian Merlot Rosé | *Very pale and evoking strawberry juice; dry enough to be drinkable* |
| £3.99 | Q | Sainsbury's Chilean Cabernet Sauvignon Rosé | *Nearly red in colour and 13% alcohol; nice drop of blackcurrant fruit with plenty of bite* |

PERSONAL NOTES:

....
....
....
....
....
....
....
....
....
....
....
....
....

£2.79	**V**	Sainsbury's Special Cuvée Liebfraumilch	*Deeply unfashionable but enjoyably grapey, balanced wine; 10% alcohol*
£2.99	(VQ)	Sainsbury's Special Cuvée Niersteiner Gutes Domtal	*Excellent, simple hock with ripeness and freshness; well made; 10% alcohol*
£2.99		Sainsbury's Special Cuvée Piesporter Michelsberg	*Slight prickle in the mouth; lively and refreshing, and only 9% alcohol*
£2.99	(VQ)	Sainsbury's Sicilian White	*Terrific, herbaceous, refreshingly dry white; charming and very cheap*
£3.49	**V**	Bereich Bernkastel Riesling 1997	*Lovely appley nose to this Moselle; good weight, low alcohol (9.5%); ghastly label*
£3.69	**V**	Chenin Blanc Vin de Pays du Jardin de la France 1998	*Loire bargain; mango and grapefruit on the nose, fresh and slurpable fruit*
£3.99		D'Istinto Catarratto Chardonnay 1998	*Australian (Hardys) inspired Sicilian dry white with tang, length and a frisky finish*
£3.99	(VQ)	Oaked Viura, Alteza 640	*Zippy, tangy, fresh, dry Spanish with sea-breeze nose and generous, brazen fruit*
£4.49		Hilltop Chardonnay 1997	*Cabbagy nose to this Hungarian number, but it has a likeable, soft heart*
£4.79		Nobilo White Cloud 1998	*Old-fashioned 'medium' New Zealand white, but with likeable, fresh balance*
£4.99		Casablanca Sauvignon Blanc 1998	*Stacks of fruit, slightly green, very crisp; from Chile*
£4.99	(VQ)	Marc Xero Chardonnay 1998	*Minerally fruit, creamy depths in this screwtop-bottled bargain from Salento, Italy*

£3.49	**V**	Alsace Blanc de Blancs 1997	*Smoky, canned-fruit nose; herby, sappy, fleshy fruit; tangy finish; cheap for Alsace*
£4.49	**Q**	Hardys Riesling Gewürztraminer 1998	*Lychee scent and steely-appley fruit in this exciting Australian blend*
£4.49	**V**	Le Trulle Chardonnay del Salento 1997	*RIchly coloured and scented and opulent fruit from Italy's torrid deep south; amazing*
£4.69		Penfolds Rawson's Retreat Bin 21 Semillon-Chardonnay 1998	*Generous Australian, dry style but heaps of lush, peachy, limey, fresh fruit*
£4.79	**VQ**	Vouvray La Couronne des Plantagenets 1998	*Demi-sec Loire classic – luscious rather than sweet; honey notes, limey finish; superb*
£4.99		Bankside Gallery Chardonnay 1997	*Vin de pays d'Oc with yellow colour, peachy nose and arty label; weighty at 13% alcohol*
£4.99	**Q**	Canepa Gewürztraminer 1998	*Chilean with grapefruit on the nose and intriguing, complex fruit; 13.5% alcohol*
£4.99		James Herrick Chardonnay 1998	*Much advertised but nevertheless satisfyingly big-fruited, balanced vin de pays d'Oc*
£4.99	**Q**	La Baume Sauvignon Blanc 1998	*The nose positively sings on this lush, fruit-laden vin de pays d'Oc*
£4.99		Lindemans Bin 65 Chardonnay 1998	*Ubiquitous but dependably generous and flavour-packed Australian*
£4.99	**VQ**	Stonybrook Vineyards Chardonnay 1997	*Californian with buttered-toast aroma and soft, sophisticated style; grand*

£5.99	**Q**	Lindemans Limestone Coast Chardonnay 1997	*Rich nose and colour to this old-fashioned, buttery Australian big-brand; 13% alcohol*

| £7.99 | Ⓠ | Penfolds Organic Chardonnay-Sauvignon Blanc 1997 | *Gold colour, melon-asparagus nose, lavish fruit; exceptionally good Oz; 13.5% alcohol* |
| £8.49 | Ⓠ | Graacher Himmelreich Riesling Spätlese, von Kesselstatt, 1997 | *Elegant, steely Moselle with thrilling racy ripeness; will convert you to German wine* |

SHERRY

| £3.95 | Ⓥ︎Ⓠ | Sainsbury's Manzanilla | *Very pale, very dry, tangy sherry of real character at an amazingly low price; drink chilled* |
| £3.49 ½ bottle | Ⓠ | Sainsbury's Premium Palo Cortado | *Superb dry sherry with nuttiness, dried-fruit aroma and flavours; drink it cool* |

Canepa Gewürztraminer from Chile is a complex, subtle, dry white to drink as an aperitif, perhaps with salty almonds or a slice of salami.

SOMERFIELD

The Bristol-based supermarket chain, now in the process of absorbing the Kwik Save stores, specialises quite brazenly in wines from the lower end of the price scale – and discounts even the cheapest lines on a regular basis. This is really the only supermarket still occasionally offering wines at £1.99, though these are always special offers for limited periods and the wines in question tend to sell out very fast. At any time, you're likely to find a few of the wines mentioned below on discount, so it's worth regularly checking out your local branch, if you have one, to see what special offers are current.

Somerfield's own-label wines, sourced by the admirable Angela Mount, are legendary. They are among the best and cheapest of all.

(V) Special for value **(Q)** Special for quality and interest **(VQ)** Special for value, quality and interest

Bright Brothers' Reserve San Juan Cabernet Sauvignon is a forceful red to drink with game.

RED LIGHT-MIDDLEWEIGHTS

£2.99	V	Somerfield Argentine Country Red 1998	*It's light in weight, but perfectly substantial and slurpable*
£2.99		Somerfield Portuguese Red 1998	*Dry, fiercely flavoursome; try with grilled sardines*
£3.29		Somerfield Vin de Pays des Coteaux de l'Ardèche	*Edgy southern French with dark, keen fruit*
£3.99		Cabernet Sauvignon delle Venezie 1998	*Italian with blackcurranty style and firm finish*
£3.99	Q	Sierra Alta Tempranillo 1998	*From La Mancha, Spain, a berryish middleweight with assertive flavour; good*
£4.29		Beaujolais AC, JP Selles 1998	*Cheap for Beaujolais; decent quaffer, though not exactly archetypal Gamay style*
£4.29	VQ	Chilean Merlot, Viña Morande 1997	*Dodgy colour but super-ripe blackberry nose and dark, lip-smacking fruit; long finish*
£4.99		Barbera d'Asti, Bricco Zanone 1997	*Cherry-bright and lushly fruity classic from Piedmont, Italy; a pasta master*

RED HEAVYWEIGHTS

£3.99	Q	Bright Bros Atlantic Vines Baga 1997	*Brambly, dark-fruit and meaty (13% alcohol) red from Portugal; big wine*
£3.99		Bright Bros Navarra Garnacha 1997	*Dark and juicy stuff from Spain's Navarra region; good grip*
£3.99		Goûts et Couleurs 1997	*A 'concept' vin de pays d'Oc, but likeable with soft, healthy fruit*
£3.99	VQ	Portada Red 1997	*Big, sweet, minty Portuguese of tremendous quality; look for the hint of honey!*
£4.25	Q	Château St Benoit Minervois 1998	*Immediately appealing, southern-French red; very clearly well made*
£4.29		South African Cabernet Sauvignon 1998	*Clean and bright with a marked cassis fruit*

£4.49		Chilean Cabernet Sauvignon 1997	*Plump, blackcurranty example with 13.5% alcohol*
£4.49		Le Trulle Primitivo Salento 1997	*Deep-south Italian; darkly toasted, slightly peppery, but soft – and 13% alcohol*
£4.79		Cono Sur Pinot Noir 1997	*Nice raspberry nose to this earthy Chilean burgundy-style red*
£4.99	Q	Bright Bros San Juan Reserve Cabernet Sauvignon 1998	*High-quality (and high-octane 14% alcohol) Cabernet from Argentina; insinuating*
£4.99	V	Hardys Stamp Series Shiraz-Cabernet Sauvignon 1998	*Solid South Australian with firm, lipsmacking cassis fruit; above-average value*
£4.99		James Herrick Cuvée Simone 1997	*Startlingly vivid colour and burstingly ripe fruit in this big-name vin de pays d'Oc*
£4.99	Q	Laperouse Syrah Cabernet 1995	*Vin de pays d'Oc by Australia's Penfolds; mature, fruity and firm; delicious*
£4.99	Q	Riparosso Montepulciano d'Abruzzo 1996	*Bumper rich, nutskin-dry-finishing Italian classic; biggest branches only*
£4.99		Trincadeira Preta 1997	*Dark, dense, oaky Portuguese; spicy, glyceriny, plummy; and, yes, good*

SPLASHING OUT ON RED WINES

| £5.49 | VQ | Redwood Trail Pinot Noir 1997 | *Elegantly delicious Californian red; cheaper at Somerfield than some outlets* |

PINK

| £3.99 | | Santa Julia Syrah Rosé 1998 | *Argentinian; strange synthetic colour but firm, reassuring fruit and dry finish; not bad* |

£2.99		Somerfield Argentine White 1998	*Surprisingly (at the price) flavoursome, soft dry white; needs plenty of chilling*
£2.99	**V**	Somerfield Sicilian White 1998	*Fresh, herbal-scented dry white; particularly good value at this price*
£2.99		Somerfield Vin de Pays Comte Tolosan 1998	*Clean and bright, from France's south*
£3.79	**Q**	Colombard 1998	*Ultra-fresh, lushly fruity South African; very easy drinking*
£3.99	**VQ**	Alsace Blanc de Blancs 1997	*From top co-op at Turckheim; fresh, lush dry with typical spice of Alsace*
£3.99	**Q**	Santa Catalina Verdejo Sauvignon 1998	*From Rueda, Spain; crisp, nettley-grassy, dry refresher with real heart*
£3.99		South African Chardonnay 1998	*Decent, bright, mouth-filling wine from Robertson region of the Cape; 13.5% alcohol*
£3.99		Two Tribes	*Anonymous but pleasantly soft, dry blend from Chile and South Africa*
£4.49		Grecanico Chardonnay 1998	*Sicilian dry white with zippy, herbaceous fruit and limey finish*
£4.99		Domaine du Bois Viognier, Vin de Pays d'Oc 1997	*Prominent melon nose and soft, insinuating fruit; lush, dry style with low acidity*
£4.99		Fiuza Bright Chardonnay 1997	*From Ribatejo, Portugal; clean Chardonnay with a distinct pineappley whiff*

WHITE HEAVYWEIGHTS

| £2.99 ½ bottle | **V** | Samos Muscat | *Bronze-coloured Greek-island stickie, very grapey and delicious – and 15% alcohol* |
| £3.29 | **V** | Somerfield Vin de Pays des Coteaux de l'Ardèche | *Crisp Rhône white with a surprising amount of plumptious weight* |

£3.99	Q	Kleinbosch Chenin Blanc Chardonnay 1998	*Heady (13.5% alcohol) South African with ripe pear and peach style*
£4.49	V	Le Trulle Chardonnay del Salento 1997	*Richly coloured and scented and opulent fruit from Italy's torrid deep south; amazing*
£4.69		Penfolds Rawson's Retreat Bin 21 Semillon-Chardonnay 1998	*Generous Australian, dry style but heaps of lush, peachy, limey, fresh fruit*
£4.99		Histonium Chardonnay 1998	*Fun label on this nicely oaked Italian; complex nose, shades of toffee apples*

FIZZ

£3.99	Moscato Spumante	*Grapey-sweet but not unrefreshing Italian frother; must be well chilled*
£4.99	Somerfield Cava Rosado	*Spanish pink fizz; dry, pleasantly fresh and strawberry scented*

SHERRY

£3.99 50cl		Gonzalez Byass Elegante	*Nicely sized bottle for this classic, pale, dry, aromatic sherry to drink chilled*
£3.99	V	Somerfield Fino Sherry	*Good-value, dry style made by famed Lustau bodega; drink it well chilled*

STOP & SHOP

see CO-OP

TANNERS

Buying wine from a long-established family wine merchant like Tanners is a pleasure in its own right. The firm is based in an impossibly picturesque Tudor-fronted courtyard building in Shrewsbury, and browsing in this place is something akin to a wine enthusiast's vision of heaven on earth. And if Shropshire is a bit distant from you, Tanners has an exceedingly efficient nationwide mail-order business. It delivers orders with reassuring speed, and at no charge (on mainland Great Britain) for orders valued at £80 upwards.

You start with Tanners' wine list. It's a 128-page colour-illustrated paperback and encompasses every kind of wine and spirit. There are a thousand wines, pages of spirits (many of them wonderfully arcane) and the biggest choice of Riedel wine glasses (the best) to be found anywhere.

True, most of the wines are priced at over £5, but there are dozens below that magic mark, and of course hundreds not far above – the sort of wines to encourage you to experiment and expand your wine-drinking brief. The wines reviewed here are the veritable tip of the Tanners iceberg.

Besides the main Wine Market in Shrewsbury, there are three other shops, at Bridgnorth (36 High Street, tel: 01746 763148), Hereford (4 St Peter's Square, tel: 01432 272044) and Welshpool (Brook Street, tel: 01938 552542).

Mail order from: Tanners Wines Ltd, 26 Wyle Cop, Shrewsbury SY1 1XD. Tel: 01743 234455. Fax: 01743 234501. E-mail: sales@tanners-wines.co.uk

V Special for value **Q** Special for quality and interest **VQ** Special for value, quality and interest

£3.85	(VQ)	Promesa Tinto 1994	*'Red-fruit' style to this Spanish middleweight with gently oaked suppleness; lovely*
£4.40	**V**	Lar de Barros Tempranillo 1997	*Unoaked Spanish (Estramadura region) red with easy charm*
£4.75		Cuvée Harmonique, Domaine la Condamine l'Evêque, 1998	*Harmonious indeed!; supple young southern-French blend with heart and fleshiness*

RED HEAVYWEIGHTS

£4.15		Marqués de Aragón Garnacha, Calatayud, 1998	*Blockbuster (14.5% alcohol) dark-fruit from Spain; bit raisiny; overripe but memorable*
£4.60		Boundary Valley Cinsaut/Cabernet Sauvignon 1998	*Well-built (13.5% alcohol) South African; stacks of hedgerow fruit; likes a bit of airing*
£4.65		Corbières, Château de Montrabech, 1996	*Mature southern-French AC, rounding out nicely with spicy, eager fruit*
£4.95		Concha Y Toro Merlot 1998	*From Chile's biggest producer, a consistently delicious, ripe, dark Merlot*
£4.95		Fitou Les Belles Roches 1997	*Typical, hearty light-heavyweight from this Midi AC, by excellent Mont Tauch co-op*

SPLASHING OUT ON RED WINES

| £6.90 | (VQ) | Lirac Les Queyrades, Domaine Méjan-Taulier, 1996 | *The juiciest, most seductive red anywhere at this price; a Rhône classic; fabulous* |
| £8.80 | | Aglianico del Vulture, Casa Vinicola d'Angelo, 1996 | *Crazy name, from Vulture volcano in southern Italy, and an unforgettable, rich, dark wine* |

PINK

| £4.95 | Château de Fonscolombe Rosé 1998 | *From Provence; onion-skin colour and floral peach-skin nose; healthy and fresh* |

WHITE LIGHT-MIDDLEWEIGHTS

£3.95	Cépage Colombard, Côtes de Gascogne, 1998	*Modern-method, crisp, dry vin de pays from good Plaimont co-op in Gascony*
£4.20	Rio Grande Chenin Blanc 1997	*Easy (just short of bland), dry white by famed Lurton flying winemaker; Argentina*
£4.30 (V)	Moselland Riesling Kabinett 1997	*Keen, appley, near-bracing, fresh Moselle from a good co-operative; try it*
£4.40	Los Vilos Chardonnay 1998	*Brisk Chilean (Concha Y Toro) with grassy nose and a crunchy, appley fruit*
£4.50 (Q)	Domaine de Rieux 1998	*Brilliantly zingy, fresh, dry white from Gascony by cult wine-armagnac makers Grassa*
£4.50	Marqués de Cáceres Blanco 1997	*Modern-style white Rioja – brisk, fresh, unserious and definitely dry*
£4.55	Frascati Superiore, Cantina Sociale di Frascati, 1997	*For when only a famous name will do; a brisk example of Rome's café wine*
£4.60	Sunnycliff Colombard/ Chardonnay 1998	*Cheerfully scented, soft-dry style with refreshing easiness from Victoria, Australia*
£4.85 (Q)	Tanners Sauvignon Vin de Pays d'Oc 1998	*Tangy and authentically grassy; 'house' Sauvignon from France's south*
£4.95	Sunnycliff Sauvignon Blanc 1998	*Nice bloom of gooseberry and keen, faintly tropical fruit; from Victoria, Australia*

WHITE HEAVYWEIGHTS

£4.25   Castillo de Liria Moscatel   *Rivettingly sweet, grapey, sticky from Valencia, Spain; drink well chilled*

£4.90   Domaines Virginie Marsanne 1998   *Perfumed vin de pays d'Oc; dry with attractive canned-fruit nuances; works well*

SPLASHING OUT ON WHITE WINES

£5.95 (VO) Jim Barry Clare Valley Unwooded Chardonnay 1997   *Greeny-gold; ripe-apple pong; pure, elegant, mineral, voluptuous Australian*

FIZZ

£4.35   Moscato Spumante La Rosa   *Shamelessly sweet and light-hearted frothing fizz from Italy; must be well chilled*

PERSONAL NOTES:

. . . .
. . . .
. . . .
. . . .
. . . .
. . . .
. . . .
. . . .
. . . .
. . . .
. . . .
. . . .
. . . .
. . . .
. . . .
. . . .
. . . .
. . . .

TESCO

Britain's biggest retailer has 566 supermarkets throughout the country and, as might be expected, an epic collection of wines. Tesco's strength is its huge range, and therein lies its weakness, too. Gawking at all these wines side by side on crowded shelves in the manic atmosphere and brutal, shadowless light of a hypermarket can make the business of effecting a rational choice of wine a very trying one indeed.

Are the wines any good? Well, yes, they are, if you can pick the good ones out from among the usual serried ranks of Blossom Hill, Gallo, Piat d'Or and other ubiquitous brands. In common with other supermarkets, Tesco do regular special offers on their wines – sweeping gestures such as 20 per cent off all wines from a particular country or region are a regular occurrence – so do take time to check out just what promotion is in force at the time you pop in.

You get a discount of 5 per cent on six or more 'units' of wine bought in one go. So if your purchase consists of any six bottles, half-bottles or wine boxes, you qualify.

While Tesco branches all carry a pretty good cross-section of the wine range, only the very biggest hypermarkets carry the lot, so there is no guarantee you'll find all the wines you're after in any given branch. To check if your local branch has the wines you want, you can ring Tesco's Customer Service line. Give them the details, including the name of the branch you intend to visit, and they can look up in their 'range matrix' whether the store in question is among those stocking the wines in question. The number is (Freephone) 0800 505555.

Ⓥ Special for value Ⓠ Special for quality and interest ⓋⓆ Special for value, quality and interest

RED LIGHT-MIDDLEWEIGHTS

£2.99	(V)	Tesco Sicilian Red	*Bright cherry colour; dry wine but with sweet notes; earthy and slurpable*
£3.79	(Q)	Picajuan Peak Sangiovese	*Juicy, vibrant young Argentine from the grape of Chianti; middleweight but 13% alcohol*
£3.99		Great with Indian 1997	*Decent, soft Chilean; advice that it's 'especially great with ... Madras' is wayward*
£3.99	(Q)	Picajuan Peak Bonarda	*Blackberries, strawberries and even bananas in this midweight but meaty Argentine slurper*
£3.99	(Q)	Santa Ines Malbec 1997	*Pale colour but delicious briary style; from Chile*
£4.49	(VQ)	Santa Ines Carmenère 1998	*Seductive Chilean with genuine suppleness of texture and cool ripeness; delish*
£4.49		Viña Mara Rioja	*Cheap for Rioja, a pure-fruit, unoaked style with youthful, fleshy vigour*
£4.49		Waimanu Dry Red	*From New Zealand; has a likeable eucalyptus, sappy-silk style to the red fruit*
£4.99		Beyers Truter Pinotage	*Brambly, juicy number from South Africa's home-grown Pinotage grape*
£4.99		Cono Sur Pinot Noir 1997	*Nice raspberry nose to this earthy Chilean burgundy-style red*
£4.99		Greenwich Meridian 2000	*Gimmicky ('The Millennium starts here') but decent 1997 Bordeaux; soft and leafy*

RED HEAVYWEIGHTS

£3.99	(Q)	Picajuan Peak Malbec 1998	*Brilliant Argentinian mouth-filler from excellent La Agricola estate; vegetarian*

£4.49		Le Trulle Primitivo Salento 1997	*Deep-south Italian; darkly toasted, slightly peppery, but soft – and 13% alcohol*
£4.79		Belafonte Baga 1998	*New vintage – the 1997 was (and is) excellent*
£4.99		Bright Bros San Juan Reserve Cabernet Sauvignon 1998	*High-quality (and high-octane 14% alcohol) Cabernet from Argentina; insinuating*
£4.99		Buzet Cuvée 44 1997	*Well-concentrated and satisfying, ripe red from an obscure SW France appellation*
£4.99	ⓋⓆ	Dão Dom Ferraz 1997	*Dark colour, dark fruit to this powerful (13% alcohol) Portuguese*
£4.99	Ⓥ	Hardys Stamp Series Shiraz-Cabernet Sauvignon 1998	*Solid South Australian with firm, lipsmacking cassis fruit; above-average value*
£4.99		James Herrick Cuvée Simone 1997	*Startlingly vivid colour and burstingly ripe fruit in this big-name vin de pays d'Oc*
£4.99		Long Mountain Cabernet Sauvignon 1997	*Pretty decent South African; ripe but not overheated; stands up to assertive food*
£4.99	Ⓥ	Terra Mater Zinfandel Syrah 1998	*Dark and dense (14% alcohol) Chilean with purity and insinuating fruit; easy drinking*
£4.99		Tesco Monster Spicy Red	*Argentine Syrah with plenty of woof (14% alcohol) but smooth rather than monstrous*
£4.99	ⓋⓆ	Undurraga Pinot Noir 1998	*Brilliant Chilean effort at reproducing the slick summer-fruit style of red burgundy*

SPLASHING OUT ON RED WINES

£7.99	Ⓠ	Norton Privada 1996	*Top Argentine oaked red from a grand mélange of Bordeaux grapes; rich and smooth*

£3.29	V	Retsina Kourtaki	*Cheap but fresh and typically pine-resiny reminder of all those taverna nights*
£3.79		Gyöngyös Estate Chardonnay 1998	*Recognisable Chardonnay with hints of cream and apple; Hungarian*
£3.99		Lenz Moser Grüner Veltliner 1997	*Brisk, dry Austrian; tangy, aromatic style with a hint of spice*
£3.99	Q	Orvieto Classico Abboccato	*Lovely, soft, off-dry Umbrian (middle Italy) classic with peaches and, maybe, macaroon*
£3.99	V	Picajuan Peak Chardonnay 1998	*Clean, characterful, refreshing Argentinian unoaked Chardonnay*
£3.99	Q	Tesco Steinweiler Kabinett	*Bright, clean and luscious quality wine from Germany's Rheinpfalz*
£3.99		Two Tribes	*Anonymous but pleasantly soft, dry blend from Chile and South Africa*
£4.99	(VQ)	Marc Xero Chardonnay 1998	*Minerally fruit, creamy depths from Italy; in this screwtop bottle*

£4.49	Q	Alsace Pinot Blanc	*Fine, spicy aroma and complex, layered fruit in this pungent classic from Alsace*
£4.49	V	Le Trulle Chardonnay del Salento 1997	*Richly coloured and scented and opulent fruit from Italy's torrid deep south; amazing*
£4.49		Picajuan Peak Viognier 1998	*Agreeably soft, nutty Argentinian with whiffs of figs and other exotic things; fun*
£4.99	Q	Gemma Gavi 1997	*Acutely, mind-concentratingly focused fruit in this Italian classic from Cortese grapes*
£4.99		James Herrick Chardonnay 1998	*Much advertised but nevertheless satisfyingly big-fruited, balanced vin de pays d'Oc*

£4.99	Lindemans Bin 65 Chardonnay 1998	*Ubiquitous but dependably generous and flavour-packed Australian*
£4.99	Luis Felipe Edwards Chardonnay 1998	*Absorbing, peachy-pineapple richness in this near-unctuous Chilean*
£4.99 (VQ) ½ bottle	Old Penola Estate Botrytis Gewürztraminer 1997	*Gorgeous, honeyed sweet Australian with lychee whiff, clean finish and 14.5% alcohol*
£4.99	Tesco Crisp Elegant White	*Intensely fruity Italian 'concept' wine with brisk, crunchy, mouth-filling fruit*
£4.99 (Q)	Tesco Pfalz Auslese	*Only 10.5% alcohol but weighty, honeyed, sweet-but-fresh German pud wine*
£4.99 (Q)	Tesco Smooth Voluptuous White	*Extravagantly buttery, oaked Australian Chardonnay 'concept' wine; great fun*
£4.99	Undurraga Gewürztraminer 1998	*Peculiarly named Chilean vineyard and European grape but exotic, intriguing wine*

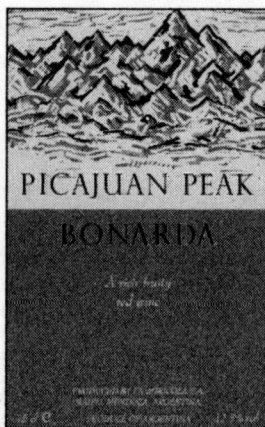

Picajuan Peak Bonarda from Argentina has enough heart to drink well with most meat dishes – even spicy ones.

| £6.99 | (VQ) | Villa Maria Sauvignon Blanc 1998 | *Blast of slaking Sauvignon from this brilliant New Zealander; keen price for a Kiwi* |
| £8.49 | (VQ) | Tim Adams Semillon 1996 | *World-class dry white of transcending quality by top Oz grower; an experience* |

SHERRY

| £3.49
½ bottle | (Q) | Tesco Superior Manzanilla Sherry | *Real, fine, dry pale sherry; smoky, tangy, pungent, refreshing; drink chilled* |
| £4.49
½ bottle | (Q) | Tesco Superior Palo Cortado Sherry | *Grand, nutty dry sherry with great style; serve slightly cool* |

PERSONAL NOTES:

. . . .
. . . .
. . . .
. . . .
. . . .
. . . .
. . . .
. . . .
. . . .
. . . .
. . . .
. . . .
. . . .
. . . .

THRESHER WINE SHOPS

see FIRST QUENCH

UNWINS

'**B**ritain's largest independent wine merchant' remains, amazingly, a family-owned company and I cannot resist mentioning that the family concerned rejoices in the name of Rotter (would I lie?). Two years ago, Unwins swallowed up the Davison off-licence chain and now has about 350 branches in London and the home counties. A few lie further afield, ranging north to Norfolk and Lincolnshire and west to Oxfordshire and Wiltshire.

These are old-fashioned shops with a big, well-chosen spread of well-known brands and a few happy surprises, even at under £5. Buy over six bottles (mixed or otherwise) and you qualify for a 10 per cent discount.

V Special for value **Q** Special for quality and interest **VQ** Special for value, quality and interest

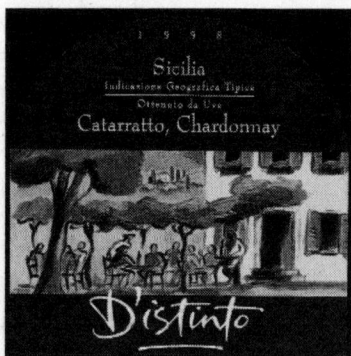

D'Istinto Catarratto-Chardonnay is a dry, tangy Sicilian white which will suit seafood from prawns to plaice.

RED LIGHT–MIDDLEWEIGHTS

| £4.49 | Viña Albali Reserva 1993 | *Oaky Spanish (Valdepeñas region) with sweet fruit and emphatic maturity* |
| £4.99 | Barbera d'Asti Le Monferine 1997 | *Bouncy, cherry-fruit but relishably dry-finishing Italian* |

RED HEAVYWEIGHTS

£3.99 (VQ)	Portada Red 1997	*Big, sweet, minty Portuguese of tremendous quality; look for the hint of honey!*
£4.49	Fitou 1997	*Solid example from the dependable winemakers co-op of Mont Tauch*
£4.49	Pedras do Monte 1998	*Plummy, juicy Portuguese with a likeable, old-fashioned robustness*
£4.99 (Q)	Belafonte Baga 1997	*Pitch-dark Portuguese with a big burst of berry fruit; weighty, good with spicy food*
£4.99	Carmen Cabernet Sauvignon 1996	*Dependable Chilean with soft, approachable fruit and a bit of backbone, too*
£4.99	James Herrick Cuvée Simone 1997	*Startlingly vivid colour and burstingly ripe fruit in this big-name vin de pays d'Oc*

PERSONAL NOTES:

. . . .
. . . .
. . . .
. . . .
. . . .
. . . .
. . . .
. . . .
. . . .
. . . .
. . . .
. . . .

WHITE LIGHT-MIDDLEWEIGHTS

£3.69	Retsina Kourtaki	*Cheap but fresh and typically pine-resiny reminder of all those taverna nights*
£3.75	Chapel Hill Oaked Chardonnay 1997	*Pretty good lightweight with appley-toasty undercurrent, from Lake Balaton, Hungary*
£3.99	D'Istinto Catarratto Chardonnay 1998	*Australian (Hardys) inspired Sicilian dry white with tang, length and a frisky finish*
£4.49	Canepa Sauvignon Blanc 1997	*Grassy-fresh, green-fruited but not over-acid Chilean Sauvignon*

WHITE HEAVYWEIGHTS

£4.69	Penfolds Rawson's Retreat Bin 21 Semillon-Chardonnay 1998	*Generous Australian, dry style but heaps of lush, peachy, limey, fresh fruit*
£4.99	James Herrick Chardonnay 1998	*Much advertised but nevertheless satisfyingly big-fruited, balanced vin de pays d'Oc*
£4.99	Lindemans Bin 65 Chardonnay 1998	*Ubiquitous but dependably generous and flavour-packed Australian*
£4.99 Q	Samos Muscat	*Ultra-sweet ambrosia from the Greek island; gorgeous pud wine or aperitif; 15% alcohol*
£4.99 Q	Viñas del Vero Chardonnay 1998	*Luxurious Spanish burgundy-style white from the Somontano region*

SPLASHING OUT ON WHITE WINES

| £5.99 | Viña Esmeralda 1997 | *Grapey-lychee-smoky, brilliantly exotic but refreshing dry white by Torres in Spain* |

VICTORIA WINE

see FIRST QUENCH

WAITROSE

The country's poshest supermarket chain, the grocery arm of the John Lewis Partnership, has 117 branches, but none is to be found farther north than Newark in the Midlands of England, or west of Dorset. Wales has to make do with one Waitrose, at Monmouth. But for the many of us not privileged to live in metropolitan suburbs or the leafier enclaves of the home counties, there is consolation – as far as wine is concerned, anyway – because you can buy all the wines sold in Waitrose by direct mail for home delivery.

After a rather difficult start, in which Waitrose Wine Direct found itself completely overwhelmed with orders and fell down badly on deliveries, there are now reassuring signs that this service is working well. (But around Christmas time, you are well advised to order before the end of November if you want your wine by 25 December.)

There are obvious advantages to using the home-delivery service. First, it gives you access to all the wines on the supermarket shelves – and a number of 'fine' wines not available in the stores. It is only in the very largest Waitrose branches that you will find all the several hundred wines on the list.

Second, provided your order is valued at £75 or more, delivery anywhere on the UK mainland is free. Note, however, that if you order mixed cases – up to 12 different wines in each dozen 'there is a small charge of £1.20 per mixed case to cover additional handling costs'.

For the latest Wine Direct list and details of special offers (they do attractive mixed-case deals six times a year and occasional bin-end sales), write to Waitrose Wine Direct, FREEPOST (SW1647), Bracknell RG12 8HX.

Tel: (Freephone) 0800 188881. Fax: (Freefax) 0800 188888.

E-mail: http://www.waitrose-direct.co.uk

Ⓥ Special for value Ⓠ Special for quality and interest ⓋⓆ Special for value, quality and interest

RED LIGHT-MIDDLEWEIGHTS

£3.49		Winter Hill Merlot/Grenache 1998	*A pale but interesting vin de pays; juicy summer fruit*
£3.59	V	Côtes du Roussillon, Jeanjean, 1998	*Bargain-priced, consistently well-made, dark and peppery southern French classic*
£3.99		Les Fontanelles Merlot-Syrah 1998	*Young and supple vin de pays d'Oc with a hint of spice*
£4.25		Saumur Les Nivières 1998	*Sappy, young-fruit classic Cabernet Franc from a famed Loire appellation*
£4.29		Ermitage, Pic St Loup, 1997	*Languedoc red from Syrah grapes making a ripe and silky style, and 13% alcohol*
£4.49	V	Espiral Tempranillo/Cabernet Sauvignon 1997	*From Somontano, Spain, an impressive blackcurrant slurper; iffy label*
£4.49		Waitrose Beaujolais 1998	*Recognisable Gamay fruit; improves in the glass*

RED HEAVYWEIGHTS

£3.99		Le Faisan Syrah-Grenache 1998	*Rhône vin de pays with heaps of sweet, brambly fruit*
£3.99	V	Viña Fuerte Garnacha 1998	*Loads of fruit in this huge (13.5% alcohol) red from Calatayud, Spain*
£4.29	Q	Concha Y Toro Merlot 1998	*From Chile's biggest producer, a consistently delicious and well-priced Merlot*
£4.49		Santa Julia Malbec-Cabernet Sauvignon 1997	*Sturdy and satisfying dark-fruit red from Argentina; vegetarian*
£4.49		Sierra Alta Cabernet Sauvignon-Malbec 1996	*Superior Argentine blend; rich leathery nose and lush fruit with plenty of grip*
£4.69		Cent'are Sicilia Rosso 1997	*Roasted, earthy style to this creamily oaked Sicilian; improves with airing*

£4.99	Ⓠ	Château Pech-Latt Corbières 1997	*Beefy, organic, oaked red from an excellent Corbières (French Midi) estate*
£4.99		Cono Sur Cabernet Sauvignon 1998	*Horrid black plastic cork but a brilliantly ripe, easy-drinking Chilean*
£4.99	Ⓥ	Hardys Stamp Series Shiraz-Cabernet Sauvignon 1998	*Solid South Australian with firm, lipsmacking cassis fruit; above-average value*

PERSONAL NOTES:

....
....
....
....
....
....
....
....
....
....

e75cl   12% vol

WAITROSE
Beaujolais
APPELLATION BEAUJOLAIS CONTRÔLÉE

Produce of France
Bottled by
Cave de Bully 69210
L'Arbresle France

BOTTLED IN FRANCE FOR WAITROSE LTD BRACKNELL BERKSHIRE UK

Waitrose Beaujolais is good of its kind, and the ideal red to drink with roast chicken.

£2.99	**V**	Waitrose Vin Blanc Sec	*It's wine, it's white, it's dry; and French; clean and without faults – and cheap*
£3.49	**V**	Clearly Blanco	*Silly name but a remarkably breezy and crisp dry white from Penedès, Spain*
£3.49		Waitrose Muscadet de Sèvre et Maine 1998	*Bracing but reasonably flavoursome example of this Loire Valley appellation*
£3.79	**Q**	Domaine de Planterieu 1998	*A vin de pays from Gascony, SW France; likeably exotic and fresh*
£3.99	**Q**	Devil's Rock Riesling 1997	*A good German; bright, keen style, not at all sweet; appley fruit, citrus edge*
£3.99		Pinot Grigio, Fiordaliso, 1998	*Superior example of this trendy Italian dry white; smoky, thirst-slaking style*
£3.99	**Q**	Penfolds Rawson's Retreat Bin 202 Riesling 1998	*Australian Riesling, nothing like German; lively, gooseberry-like zing*
£3.99		Verdicchio dei Castelli di Jesi Classico, Moncaro, 1998	*Pale and shy on nose but nice minerally-nutty style; hint of spice at the finish; Italian*
£4.29		Orvieto Classico Secco, Cardeto, 1997	*Dry but interestingly herbaceous and appealing Italian from Umbria*
£4.49		Waitrose Muscadet de Sèvre et Maine sur Lie 1998	*Refreshing Loire white with more fruit than most Muscadets at this price*
£4.85		Cortese DOC, Alasia, 1997	*Very pale but with startlingly grassy, keenly refreshing style; Piedmont, Italy*
£4.99	**Q**	Bernkasteler Kuesner Weisenstein Riesling Kabinett 1995	*'Racy, nervous acidity and tingling lemon and lime fruit,' says Waitrose; it's true!*

£3.99		Deer Leap Gewürztraminer 1997	*Exotic, soft and spicy white from Hungary*
£3.99	(V)	Robert's Rock Chardonnay 1998	*Surprisingly rich, creamy style to this inexpensive South African; 13% alcohol*
£4.95	(VQ)	Alsace Pinot Blanc, Blanck, 1997	*Exotic, near-pungent, lush dry white from a top producer in France's Alsace region*
£4.99	(Q)	Cuckoo Hill Chardonnay/Viognier 1998	*Reliable vin de pays d'Oc; lush blend of peachy, honey and figgy flavours yet freshly dry*
£4.99		Domaine Boyar Barrel Fermented Chardonnay 1997	*Rich colour and coconutty-creamy style to this luxury Bulgarian*
£4.99		James Herrick Chardonnay 1998	*Much advertised but nevertheless satisfyingly big-fruited, balanced vin de pays d'Oc*
£4.99 ½ bottle		Muscat de Beaumes-de-Venise	*Magnificent, grapey-honeyed example of this sweet but uncloying Rhône classic*
£4.99	(Q)	Viñas del Vero Chardonnay 1998	*Luxurious Spanish burgundy-style white*

PERSONAL NOTES:

....
....
....
....
....
....
....
....
....
....
....
....
....
....

WINE CELLAR

These are the flagship shops of Parisa, the Warrington-based off-licence group with around 500 branches around the country. Other outlets of this company, selling wines from a core list, are Booze Buster 'no-frills, cut-price off-licences with shopfronts'. There are more than 200 of these, along with 70 Right Choice branches, which are 'local convenience shops offering a full range of fresh and packaged groceries as well as drinks, news and video rental' in the Midlands and north of England. Cellar 5 shops are local off-licences, and Berkeley Wines are 'helpful and easy wine stores which have a varied and popular selection of wines. They offer a traditional and comfortable environment and a high level of personal service.'

The most remarkable service aspect of Parisa is its move into selling wines both on- and off-licence in the same premises. In its Parisa Café Bars and Wine Cellar shops-with-café-attached, you can buy wine from the shelf at shop price and for £3 corkage drink it in the café area – where they also serve freshly baked baguettes and patisseries as well as coffee. Parisa Café Bars have a choice of about 250 wines, and Wine Cellar branches around 700.

This is an innovative company, striving to compete with the impersonal blandness of the supermarkets and to make shopping for wine as much of a pleasure as it deserves to be. If you don't have a local branch, you can order by phone on (Freephone) 0800 838251 for home delivery, or visit their website (which gives current list, prices and special-offer details) at http://www.winecellar.co.uk

Ⓥ Special for value ● Special for quality and interest ⓋⓆ Special for value, quality and interest

RED LIGHT-MIDDLEWEIGHTS

£3.99	**V**	Carta Vieja Merlot 1998	*Rather a supple and poised wine at this price; from Chile*
£3.99	**Q**	Libertad Malbec/Sangiovese 1997	*Argentinian with silkiness and lush cherry-blackberry fruit; exciting stuff*
£3.99		Pedras do Monte 1997	*Portuguese with concentration and juicy fruit; costs less elsewhere*
£3.99	**V**	Terra Boa, Tras-Os-Montes, 1997	*Portuguese middleweight with lively, grippy, mature fruit; very good value*
£4.49		Montepulciano d'Abruzzo, Umani Ronchi, 1996	*Agreeable, jolly, purply cherry-fruit from Adriatic Italy; nice clean finish*

RED HEAVYWEIGHTS

£3.99	**VQ**	Belafonte Baga 1997	*Pitch-dark Portuguese with a big burst of berry fruit; good price here*
£3.99	**VQ**	Portada Red 1997	*Big, sweet, minty Portuguese of tremendous quality; look for the hint of honey!*
£3.99	**V**	Rafael Estate Tempranillo 1997	*Argentine with depths of strawberry-blackberry fruit; airing in the glass improves it*
£4.49		La Palma Cabernet-Merlot 1997	*Rich colour, plenty of cassis and plum fruit, hint of tobacco; from Rapel, Chile*
£4.49	**V**	Santa Julia Pinot Noir 1997	*Very good attempt at the strawberry ripeness of Pinot by La Agricola, Argentina*
£4.59		Dão Tinto, Caves Alianca, 1996	*Firm, earthy Portuguese with nice dark berry-fruit flavour in the background*
£4.79		Long Mountain Cabernet Sauvignon 1997	*Pretty decent South African; ripe but not overheated; stands up to assertive food*

£4.79		Le Trulle Primitivo Salento 1997	*Deep-south Italian; darkly toasted, slightly peppery, but soft – and 13% alcohol*
£4.99		Cent'are Sicilia Rosso 1997	*Roasted, earthy style to this creamily oaked Sicilian; improves with airing*
£4.99		Chasse du Pape Côtes du Rhône 1996	*Robust red with Châteauneuf-du-Pape pretensions and even a vague similarity*
£4.99		Enate Tinto 1995	*Maturing Cabernet/Merlot from Spain with softening tannin and emerging fruit*
£4.99	Ⓥ	Hardys Stamp Series Shiraz-Cabernet Sauvignon 1998	*Solid South Australian with firm, lipsmacking cassis fruit; above-average value*
£4.99	Ⓠ	Marqués de Griñon Rioja Crianza 1996	*Quality Rioja under £5 is rare and this oaky, squishily fruity, mature one is rare indeed*
£4.99	ⓋⓆ	Norton Barbera 1997	*Chunky (14% alcohol) Argentinian with spice and wild-fruit flavours; gripping stuff*
£4.99		Norton Cabernet Sauvignon 1997	*Satisfying, ripe blackcurrant style from a dependable estate in Mendoza, Argentina*
£4.99		Norton Malbec 1997	*Grippy, ripe and soft red from renowned estate in Mendoza, Argentina*
£4.99	Ⓠ	Viñas del Vero Cabernet Sauvignon 1996	*Silky, oaked Spanish with smooth cassis fruit and velvety length; can't be much left*

SPLASHING OUT ON RED WINES

£5.49	ⓋⓆ	Redwood Trail Pinot Noir 1997	*Elegantly delicious Californian red; cheaper here than some places*
£7.99		Norton Privada 1996	*Top Argentine oaked red from a grand mélange of Bordeaux grapes; rich and smooth*

WHITE LIGHTWEIGHTS

£3.79		Chapel Hill Sauvignon Blanc 1998	*Respectable, brisk Sauvignon from Hungary at a keen price*
£3.99		Carta Vieja Sauvignon Blanc 1998	*Recognisable green, sea-breeze Sauvignon style at a fair price from Chile*
£3.99		D'Istinto Catarratto Chardonnay 1998	*Australian (Hardys) inspired Sicilian dry white with tang, length and a frisky finish*
£3.99	Q	Penfolds Rawson's Retreat Bin 202 Riesling 1998	*Australian Riesling, nothing like German; lively, gooseberry-like zing*
£4.99		La Baume Sauvignon Blanc 1998	*The nose positively sings on this lush, fruit-laden vin de pays d'Oc*
£4.99		Wild Trout 1997	*Vin de pays d'Oc with freshness and a crispness; get the 1998 if they have it*

WHITE HEAVYWEIGHTS

£4.49	V	Hardys Stamp Series Chardonnay-Semillon 1998	*Good price for this emphatically flavoured Australian*
£4.69		Penfolds Rawson's Retreat Bin 21 Semillon-Chardonnay 1998	*Generous Australian, dry style but heaps of lush, peachy, limey, fresh fruit*
£4.99	VQ	Cordillera Estate Chardonnay Reserva 1997	*Lavishly coloured wine from Casablanca, Chile, with purity and elegance; special*
£4.99		James Herrick Chardonnay 1998	*Much advertised but nevertheless satisfyingly big-fruited, balanced vin de pays d'Oc*
£4.99		Lindemans Bin 65 Chardonnay 1998	*Ubiquitous but dependably generous and flavour-packed Australian*

| £4.99 | Ⓠ | Viñas del Vero Chardonnay 1997 | *Maturing, buttery-rich, pineapple and peaches Spanish (Somontano) 'burgundy'* |

SPLASHING OUT ON WHITE WINES

£6.99	ⓋⓆ	Château de Berbec AC 1er Côtes de Bordeaux 1992	*Gorgeous, nectareous, honeyed but clean-finishing 'pudding wine' at a rare price*
£6.99	ⓋⓆ	Villa Maria Sauvignon Blanc 1998	*Blast of slaking Sauvignon from this brilliant New Zealander; keen price for a Kiwi*
£7.99	Ⓠ	Bonterra Chardonnay 1996	*Heaven in a bottle from a famed organic estate in California; pure, lavish Chardonnay*

FIZZ

| £4.99 | | Asti Spumante Botticella | *Standard but very decent sweet frother from Asti in Italy; needs to be cold* |
| £4.99 | | Marqués de Monistrol Cava Brut | *Big-name sparkler from Penedès, Spain; dry but fruity 'champagne'* |

PERSONAL NOTES:

....
....
....
....
....
....
....
....
....
....

WINE RACK

see FIRST QUENCH

A brief vocabulary of wine

The purpose of this little dictionary is to explain the significance and meaning of some of the less familiar words used to describe the wines in this book. These may be terms from labels, grape varieties or words used in the notes made about the wines.

abboccato – Medium-dry white wine style of Italy.

AC – *see* **Appellation d'Origine Contrôlée.**

acidity – To be any good, every wine must have the right level of acidity. It gives wine the element of dryness or sharpness it needs to prevent cloying sweetness or dull wateriness. Too much acidity, and wine tastes raw or acetic (vinegary). Winemakers strive to create balanced acidity – either by cleverly controlling the natural processes, or by adding sugar and acid to correct imbalances.

aftertaste – The taste remaining in the mouth after you have swallowed the wine.

Aglianico – Black grape variety of southern Italy, it is romantically associated. When the ancient Greeks first colonised Italy in the seventh century BC with the prime purpose of planting it as a vineyard (Greek name for Italy was *Oenotria* – land of cultivated vines), the name for the vines the Greeks brought with them was Ellenico (as in *Hellas,* Greece), from which Aglianico is the modern rendering. To return to the point, these ancient vines, especially in the arid volcanic landscapes of Basilicata, produce excellent dark, earthy and highly distinctive wines. A name to look out for.

Alsace – France's eastern-most wine-producing region lies between the Vosges mountains and the Rhine river. Conditions there make for the production of some of the world's most delicious and fascinating white wines, always sold under the name of their constituent grapes. **Pinot Blanc** is the most affordable – and is well worth seeking out.

amontillado – *see* **sherry.**

AOC – *see* **Appellation d'Origine Contrôlée.**

Appellation d'Origine Contrôlée – Commonly abbreviated to AC or AOC, this is the system under which quality wines are defined in France. About a third of the country's vast annual output qualifies, and there are more than 400 distinct AC zones. The declaration of an AC on the label signifies the wine meets standards concerning location of vineyards and wineries, grape varieties and limits on harvest per hectare, methods of cultivation and vinification and alcohol content. Wines are inspected and tasted by state-appointed committees. The only factor an AC cannot guarantee about a wine is that you will like it – but it certainly improves the chances.

Apulia – *see* **Puglia.**

Asti – Town and major winemaking centre in Piedmont, Italy. The sparkling *(spumante)* sweet wines are inexpensive and often delicious. Typical alcohol level is a modest 7 per cent.

attack – The first impression of the wine in the mouth.

auslese – *see* **QmP.**

backbone – In my personal wine-tasting terminology, the impression given by a well-made wine in which the flavours are a pleasure to savour at all three stages: first sensation in the mouth, while being held in the mouth, and in the aftertaste when the wine has been swallowed or spat out. Such a wine is thus connected together by a 'backbone'.

Baga – Black grape variety of Portugal. Makes famously concentrated, juicy reds.

balance – A big word in the vocabulary of wine tasting. Respectable wine must get two things right: lots of fruitiness from the sweet grape juice, and plenty of acidity so the sweetness is 'balanced' with the crispness familiar in good dry whites and the dryness that marks out good reds. Some wines are noticeably 'well balanced' in that they have memorable fruitiness and the clean, satisfying 'finish' (last flavour in the mouth) that ideal acidity imparts.

Barbera – Black grape variety originally of Piedmont in Italy. Most commonly seen as Barbera d'Asti, the vigorously fruity red wine made around **Asti** – which is better known for sweet sparkling Asti Spumante. Barbera grapes are now being grown in South America, producing a sleeker, smoother style than at home in Italy.

Beaujolais – Unique red wines from the southern reaches of Burgundy, France, are made from Gamay grapes. Beaujolais Nouveau, the new wine of each harvest, is released on the third Thursday of every November to much ballyhoo.

Much of it is usually within a £5 budget and provides a friendly introduction to this deliciously bouncy, fleshily fruity wine style. The bad news is that decent Beaujolais for enjoying during the rest of the year has lately become rather expensive. If splashing out, go for Beaujolais Villages, from the region's better, northern vineyards.

Beaumes de Venise – Village near Châteauneuf-du-Pape in France's Rhône valley famous for sweet and alcoholic wine from **Muscat** grapes. Some supermarkets (such as Waitrose) have these magnificent wines in half bottles at under £5, and they are a treat not to be missed.

bianco – White wine, Italy.

bite – The impression on the palate of a wine with plenty of **acidity** and possibly **tannin**.

blanc – White wine, France.

blanc de blancs – White wine from white grapes, France. Seems to be stating the obvious, but some white wines (such as champagne) are made, partially or entirely, from black grapes.

blanc de noirs – White wine from black grapes, France.

blanco – White wine, Spain and Portugal.

bodega – In Spain, a wine producer or wine shop.

Bonarda – Black grape variety of northern Italy. Now more widely planted in Argentina, where it makes rather elegant red wines.

bouncy – The feel in the mouth of a red wine with young, juicy fruitiness. Good Beaujolais is bouncy.

bouquet – *see* **nose**.

briary – In wine tasting, associated with the flavours of fruit from prickly bushes such as blackberries.

brut – Driest style of sparkling wine. Originally a French term, for very dry champagnes developed for the British market, but now used for sparkling wines from all round the world.

Cabernet Franc – Black grape variety originally of France. It makes the light-bodied and keenly edged red wines of the Loire Valley – such as Chinon and Saumur – and it is much grown in Bordeaux, especially in the *grand* *appellation* of St Emilion. Also now planted in Argentina, Australia and North America. Wines, especially in the Loire, are characterised by a leafy, sappy style and bold fruitiness. They are best enjoyed young.

Cabernet Sauvignon – Black (or, rather, blue) grape variety now grown in virtually every wine-producing nation on earth. When ripe, the grapes are smaller than many other varieties and have particularly thick skins. This means that when pressed, Cabernet grapes have a high proportion of skin to juice – and that makes for wine with lots of colour and **tannin**. In Bordeaux, the grape's traditional home, Cabernet-based wines have always been known as *vins de garde* (wines to keep) because they take years, even decades, to evolve as the effect of all that skin extraction preserves the fruit all the way to magnificent maturity. But in today's impatient world, these grapes are exploited in modern winemaking techniques to produce the sublime flavours of mature Cabernet without having to hang around for lengthy periods awaiting maturation. While there's nothing like a fine, ten-year-old claret (and nothing quite as expensive) there are many excellent Cabernets from around the world that amply illustrate this grape's characteristics. Classic smells and flavours include blackcurrants, cedar wood, chocolate, tobacco – and even violets.

Cahors – An AC of the Lot valley in south-west France once famous for 'black wine'. This was a curious concoction of straightforward wine mixed with soupy stuff made by boiling up the new-pressed juice to concentrate it (by evaporating the water content) before fermentation. It probably tasted horrible, and long ago disappeared. Ironically, many Cahors wines today, although still largely made with the original Malbec grapes, are disappointingly light.

cantina sociale – *see* **co-op.**

Carmenère – Black grape variety once widely grown in Bordeaux but abandoned due to cultivation problems. Lately revived in South America where it is producing fine wines.

cassis – As a tasting note, signifies a wine with a noticeable blackcurrant-concentrate flavour or smell.

Catarratto – White grape variety of Sicily. In skilled hands it makes keen, green-fruit dry whites. Also used for marsala.

Cava – the sparkling wine of Spain. Most is made in the Penedès region inland from Barcelona. Although much cava is very reasonably priced, don't assume it's greatly inferior to the very much more expensive wines of Champagne. Cava is made by the same method as champagne (second fermentation in bottle) and competes very well for value.

cépage – Grape variety, French. Cépage Merlot on a label simply means the wine is made from Merlot grapes.

Chablis – Northern-most AC of France's Burgundy region. Its dry white wines from Chardonnay grapes are known for their fresh and minerally style, but the best wines also age very gracefully into complex classics. Unfortunately, the wines are uniformly expensive.

Chardonnay – The world's most popular grape variety, said to originate from the village of Chardonnay in the Mâconnais region of southern Burgundy, and now planted in every wine-producing nation. Wines are characterised by generous colour and sweet-apple smell, but styles range from lean and sharp to opulently rich.

Chenin Blanc – White grape variety of the Loire Valley, France. Now also grown farther afield, especially in South Africa. Makes dry, soft white wines and also rich, sweet styles.

cherry – In wine tasting, either a pale red colour or, more commonly, a smell or flavour akin to the sun-warmed, bursting sweet ripeness of cherries. Many Italian wines – from lightweights such as Bardolino and **Valpolicella** to serious Chianti – have this character. 'Black cherry' as a description is often used of **Merlot** wines – meaning they are sweet but have a firmness associated with the thicker skins of black cherries.

Cinsault – Black grape variety of southern France, where it is invariably blended with others in all qualities ranging from vin de pays to the pricy reds of Châteauneuf-du-Pape. Also much planted in South Africa. Effect in wine is to add keen aromas (sometimes compared with turpentine!) and softness to the blend. The name is often spelt Cinsaut.

claret – The red wine of Bordeaux, France.

clarete – On Spanish labels indicates a pale-coloured red wine. **Tinto** signifies a deeper colour.

classic – An overused term in every respect – wine descriptions being no exception. In this book, the word is used to describe a very good wine of its type, so, a 'classic' Cabernet Sauvignon is one that is recognisably and admirably characteristic of that grape.

Classico – Under Italy's wine laws, this word appended to the name of a DOC zone has an important significance. The Classico wines of the region can only be made from vineyards lying in the best-rated areas, and wines thus labelled (such as Chianti Classico, Soave Classico, Valpolicella Classico) can be reliably counted on to be a cut above the rest.

Colombard – White grape variety of southern France. Once employed almost entirely for making the wine that is distilled for armagnac and cognac brandies, but lately restored to varietal prominence in the vin de pays des Côtes de Gascogne where hi-tech wineries turn it into a fresh and crisp, if unchallenging, dry wine at a budget price. But beware, cheap Colombard (especially from South Africa) can still be very dull.

concept – A marketing term now frequently applied to wines. Examples are bottles labelled not with their national origin or grape variety, but with the name of an occasion or food with which the wine is perceived to be an apt choice, such as Tesco's Great with Indian.

co-op – Very many of France's good-quality, inexpensive wines are made by co-operatives. These are wine-producing factories whose members, and joint-owners, are local *vignerons* (vine-growers). Each year they sell their harvests to the co-op for turning into branded wines. In Italy, co-op wines can be identified by the words *cantina sociale* on the label.

Corbières – A name to look out for. It's an AC of France's Midi (deep south) and produces countless robust reds, often at bargain prices.

Cortese – Obscure white grape variety of Piedmont, Italy. At its best, makes amazingly delicious, keenly brisk and fascinating wines. Worth seeking out, even at under £5.

Costières de Nîmes – An AC of Languedoc-Roussillon in southern France. This is a name to look out for. The best red wines are marked by a notable concentration both of colour and fruit, with the earthy-spiciness of the better Rhône wines and often a pleasing hint of liquorice. Exciting red wines, sometimes reasonably priced. There are some good rosé wines, but they are hardly seen in Britain.

côte – In French, it simply means a side, or slope, of a hill. The implication in wine terms is that the grapes come from a vineyard ideally situated for maximum sunlight, good drainage and the unique soil conditions prevailing on the hill in question. It's fair enough to claim that vines grown on slopes get more sunlight than those grown on the flat, but there is no guarantee whatsover that any wine labelled 'Côtes du' this or that have been grown on a hillside anyway. Côtes du Rhône wines are a case in point. Many come from vineyards as flat as snooker tables. The quality factor is determined by the weather and the talents of the winemaker.

crianza – Means 'nursery' in Spanish. On Rioja wines, the designation signifies a wine that has been nursed through a maturing period of at least a year in oak casks and a further six months in bottle before being released for sale.

cru – A word that crops up in all sorts of contexts on French wine labels. It means 'growth' and signifies that the wine concerned is from a very particular vineyard. Under the Appellation Contrôlée laws, countless *crus* are classified in various hierarchical ranks. Hundreds of individual vineyards are described as *premier cru* or *grand cru* in the classic wine regions of Alsace, Bordeaux, Burgundy and Champagne. The common denominator in all these is that the wine can be counted on to be enormously expensive. On humbler wines, the use of the word *cru* tends to be merely for decoration.

cuvée – French for the wine in a *cuve* or vat. The word is much used on labels to imply that the wine is from just one vat, and thus of unique, unblended character. *Première cuvée* is supposedly the best wine from a given pressing because the grapes have had only the initial, gentle squashing to extract the free-run juice. Subsequent *cuvées* will have been from harsher pressings, grinding the grape pulp to extract the last drop of juice. Take all this with a pinch of salt.

demi-sec – 'Half-dry' style of French (and some other) wines. Beware. It can mean anything from off-dry to cloyingly sweet.

DOC – Stands for *Denominazione di Origine Controllata*, Italy's equivalent of France's AC. But to suggest that DOC wines can be relied upon to be Italy's best is to oversimplify. DOC just means the wine is made according to the stipulations of its zone of origin – not that it will be wonderful wine. DOCG stands for the aforementioned with e *Garantita* appended. There are 16 DOCG zones making, in theory, the best wines in Italy. Don't count on it!

earthy – A tricky word in the wine vocabulary. In this book, its use is meant to be complimentary. It indicates that the wine somehow suggests the soil the grapes were grown in, even (perhaps a shade too poetically) the landscape in which the vineyards stand. The amazing-value red wines of the torrid, volcanic southernmost regions of Italy are often described as earthy. This is an association with the pleasantly 'scorched' back-flavour in wines made from the ultra-ripe harvests of this near-sub-tropical part of the world.

edge – A wine with an edge is one with obvious (but not necessarily excessive) acidity.

élevé – 'Brought up' in French. Much used on French wine labels where the wine has been matured (brought up) in oak barrels (*élevé en fûts de chêne*) to give it extra dimensions.

Faugères – An **AC** of the Languedoc in south-west France. Source of many hearty, economic reds.

finish – The last flavour which lingers in the mouth.

fino – Pale and very dry style of sherry. You drink it thoroughly chilled – and you don't keep it any longer after opening than other dry white wines. Needs to be fresh to be at its best. *See sherry.*

flabby – Wine-speak for watery.

Gamay – The black grape that makes all red Beaujolais. It is a pretty safe rule to avoid Gamay wines from any other region. It's a grape that does not travel well.

Gewürztraminer – One of the great grape varieties of Alsace, France. At their best, the wines are perfumed with lychees and richly, spicily fruity and yet quite dry. Gewürztraminer from Alsace is, unhappily, invariably expensive but the grape is also grown with some success in Italy, South America and elsewhere, and sold at more approachable prices.

green – In flavour, a wine that is unripe, even raw. But greenness that balances what would otherwise be an excessively sweet flavour is a merit, not a flaw.

grip – In wine-tasting terminology, the sensation in the mouth produced by a wine which has a healthy quantity of **tannin** in it. A wine with grip is a good wine. A wine with too much **tannin,** or which is still too young (the tannin hasn't 'softened' with age) is not described as having grip, but as mouth-puckering – or simply undrinkable.

Grüner Veltliner – The 'national' white-wine grape of Austria. It used to make exclusively soft, German-style everyday wines, but now is behind some excellent dry styles, too.

halbtrocken – 'Half-dry' in Germany's wine vocabulary. A reassurance that the wine is not some ghastly sugared Liebfraumilch-style confection.

hock – The wine of Germany's Rhine river valleys. It comes in brown bottles, as distinct from the wine of the Mosel river valleys – which comes in green ones.

Indicazione Geografica Tipica – Italy's recently instituted wine-quality designation, broadly equivalent to France's **vin de pays.** The label has to state the geographical location of the vineyard and will often (but not always) state the principal grape varieties from which the wine is made.

insinuating – A personal tasting note which describes a wine with an appeal that grows on me, not always for easily definable reasons.

Kabinett – Under Germany's bewildering wine-quality rules, this is a classification of a top-quality (see **QmP**) wine. Expect a keen, dry style. The name comes from the cabinet or cupboard in which winemakers traditionally kept their most treasured bottles.

Languedoc-Roussillon – Vast area of southern France, including the country's south-west Mediterranean region. The source, now, of many great-value wines from countless **ACs** and **vin de pays** zones.

legs – The wine left clinging to the inside of the glass after the wine has been swirled round, then left to subside. *See also* **weight**.

length – When the pleasure-giving flavours of a wine persist in the mouth for an appreciable period after you've swallowed it, the wine is described as having length.

Malbec – Black grape variety grown on a small scale in Bordeaux, and the mainstay of the wines of Cahors in France's Dordogne region. Now much better known for producing big butch reds in Argentina.

Manzanilla – Pale, very dry sherry of Sanlucar de Barrameida, a grungy seaport on the southern-most coast of Spain. Manzanilla is proud to be distinct from the pale, very dry **fino** sherry of the main producing town of Jerez de la Frontera down the coast. Drink it chilled, and fresh – it goes downhill in an opened bottle after just a few days even if kept (as it should be) in the fridge. *See* **sherry**.

Marsanne – White grape variety of the northern Rhône Valley and, increasingly, of the wider south of France. It's known for making well-coloured wines with heady aroma and fruit.

meaty – Weighty, rich red wine style.

Merlot – One of the great black wine grapes of Bordeaux and now grown all over the world. Characteristics of Merlot-based wines attract descriptions such as 'plummy', 'plump' and 'cherry-ripe'. The grapes are larger than most, and thus have less skin in proportion to their whole. This means the resulting wines have less **tannin** than wines from smaller-berry varieties such as Cabernet Sauvignon, and are therefore more suitable for drinking while still young.

middle palate – In wine tasting, the taste as you 'chew' the wine.

Midi – Catch-all term for the deep south of France, west of the Rhône Valley.

minerally – This is a useful wine-tasting term that's rather hard to pin down. It is used for wines that are immediately likeable for their crispness, cleanness and complexity.

Minervois – AC for mostly red wines in Midi, France. Often very good value.

Monastrell – Black grape variety of Spain, widely planted in Mediterranean regions for inexpensive wines notable for their high alcohol and toughness – though they can mature into excellent, soft reds. The variety is known in France as Mourvèdre and in Australia as Mataro.

Montepulciano – Black grape variety of Italy. Best-known in Montepulciano d'Abruzzo, the juicy, purply-black and bramble-fruited red of the Abruzzi region mid-way on Italy's Adriatic side. Not to be confused with the hill town in Tuscany famous for expensive Vino Nobile di Montepulciano wine.

Moselle – The wine of Germany's Mosel River valleys. Always comes in green bottles.

Moscato – *see* Muscat.

Mourvèdre – Widely planted black grape variety of southern France. It's an ingredient in many of the wines of Provence, the Rhône and Languedoc, including the ubiquitous vins de pays d'Oc. It's a hot-climate vine and the wine is usually blended with other varieties to give sweet aromas and body to the mix. Known as Mataro in Australia and Monastrell in Spain.

Muscat – Grape variety with origins in ancient Greece, and still grown widely for the production of sweet white wines. Muscats are the wines that taste more like grape juice than any other – but the very high sugar levels ensure they are also among the most alcoholic of wines, too. Known as Moscato in Italy, the grape is much used for making sweet sparkling wines, as in Asti Spumante or Moscato d'Asti.

Navarra – DO (Denominacion de Origen) wine-producing region of northern Spain adjacent to, and overshadowed by, Rioja. Navarra's wines can be startlingly akin to their neighbouring rivals, and sometimes rather better value for money.

nose – In the vocabulary of the wine-taster, the 'nose' is the scent of a wine. Sounds a bit dotty, but makes a sensible-enough alternative to the rather bald 'smell'. The use of the word 'perfume' implies that the wine smells particularly good. 'Aroma' is used specifically to describe a wine that smells as it should, as in 'this burgundy has the authentic strawberry-raspberry aroma of Pinot Noir' (*see* **Pinot Noir**). 'Bouquet' has rather gone out of use.

note – It's pretentious to compare wine with music of course, but it's hard to resist saying that the smells and flavours of some wines seem to be placed here and there on a scale akin to musical notation.

oloroso – see sherry.

palo cortado – see sherry.

pebbly – A flavour style in dry white wine where the fresh coolness imparted in the mouth evokes the purity and simplicity of shiny pebbles. No association with seawater is intended, though.

perfume – see nose.

Periquita – Black grape variety of southern Portugal. Makes rather exotic spicy reds. The name means 'parrot'.

petrolly – When white wines from some grape varieties, especially Riesling, reach a certain age they can take on a spirity aroma reminiscent of petrol, sometimes diesel. In grand old German wines, this is considered a thoroughly good thing.

Pinotage – South Africa's own black grape variety. Makes red wines ranging from light and juicy to dark, strong and long-lived. It's a cross between **Pinot Noir** and a grape the South Africans used to call the Hermitage (thus the portmanteau name) but turns out to have been the **Cinsault**.

Pinot Blanc – White grape variety principally of Alsace, France. Florally perfumed, exotically fruity dry white wines.

Pinot Grigio – White grape variety of northern Italy. Wines bearing its name have become fashionable in recent times. Good wines have an interesting smoky-pungent aroma and keen, slaking fruit. But most are dull. Originally a French grape, there known as Pinot Gris, which is renowned for making lushly exotic – and expensive – white wines in the Alsace region.

Pinot Noir – The great black grape of Burgundy, France. It makes all the region's fabulously expensive red wines. Notoriously difficult to grow in warmer climates, it is nevertheless cultivated by countless intrepid winemakers in the New World intent on reproducing the magic appeal of red burgundy. California and New Zealand have come closest, but rarely at prices much below those for the real thing. Some Chilean and Romanian Pinot Noirs cost under £5 and are worth trying.

Primitivo – Black grape variety of southern Italy, especially the region of **Puglia**. The wines are typically dense and dark in colour with plenty of alcohol, and an earthy, spicy style. Often a real bargain. Said to be the same variety as California's Zinfandel, which makes purple, brambly wines of a very different hue.

Prosecco – White grape variety of Italy's Veneto region which gives its name to a light, sparkling wine that is much appreciated locally, but little exported.

Puglia (also **Apulia**)– The region occupying the 'heel' of southern Italy, and one of the world's fastest-improving sources of inexpensive wines. Modern wine-making techniques and large regional grants from the EU are responsible.

QbA – On a German wine label stands for *Qualitätswein bestimmter Anbaugebiet*. It means 'quality wine from designated areas' and implies that the wine is made from grapes with a minimum level of ripeness, but it's by no means a guarantee of exciting quality. Only wines labelled **QmP** can be depended upon to be special.

QmP – On a German wine label stands for *Qualitätswein mit Prädikat*. These are the serious wines of Germany, made without the addition of sugar to 'improve' them. To qualify for QmP status, the grapes must reach a level of ripeness as measured on a sweetness scale – all according to Germany's fiendish wine-quality regulations. Wines from grapes which reach the minimum level of sweetness qualify for the description of *Kabinett*. The next level up earns the rank of *Spätlese*, meaning 'late-picked'. Kabinett wines can be expected to be dry and brisk in style, and Spätlese wines a little bit riper and fuller. The next grade up, *Auslese*, meaning 'selected harvest', indicates a wine made from super-ripe grapes; it will be golden in colour and honeyed in flavour. At under £5 there is a reasonable choice of Kabinett and occasionally Spätlese. Auslese wines are more expensive, but some supermarkets (including Tesco) have budget own-label examples.

Quinta – Portuguese for farm or estate. It is pronounced KEEN-ta.

racy – Evocative wine-tasting description for wine that thrills the tastebuds with a rush of exciting sensations. Very good Rieslings often qualify.

Reserva – In Portugal and Spain, this has genuine significance. The Portuguese use it for special wines with a higher alcohol level and longer ageing, although the precise periods vary between regions. In Spain, especially in the Navarra and Rioja regions, it means the wine must have had at least a year in oak and two in bottle before release.

réserve – On French or other wines, this implies special quality, but has no meaning in official terms.

Ribatejo – Emerging wine region of Portugal. Worth looking out for on labels of red wines.

Riesling – The noble grape variety of Germany. It is correctly pronounced REEZ-ling, not RICE-ling. Once notorious as the grape behind all those boring 'medium' Liebfraumilches and Niersteiners, this grape has had a bad press. In fact, there's little if any Riesling in Germany's cheap-and-nasty plonks. But the country's best wines, the so-called *Qualitätswein mit Prädikat* grades (*see* QmP), are made almost exclusively with Riesling. These wines range from crisply fresh and appley styles to extravagantly fruity, honeyed wines from late-harvested grapes. Excellent Riesling wines are also made in Alsace and now in Australia.

Rioja – The principal fine-wine region of Spain, in the country's north-east. The pricier wines are noted for their vanilla-pod richness from long ageing in oak casks. Younger wines, labelled 'sin-crianza' (meaning they are not barrel aged – *see* **crianza**), are cheaper and can make relishable drinking.

Riserva – In Italy, a wine made only in the best vintages, and allowed longer ageing in cask and bottle.

rosso – Red wine, Italy.

Sangiovese – The local grape variety of Tuscany, Italy. It is the principal grape used for **Chianti** and is now widely planted in Latin America – often making delicious, Chianti-like wines with characteristic cherryish-but-deeply-ripe fruit and dry, clean finish. Chianti wines have become expensive in recent years and cheaper Italian wines, such as those called Sangiovese di Toscana, make a consoling substitute.

Sauvignon Blanc – White grape variety now grown worldwide. The wines are characterised by aromas of gooseberry, fresh-cut grass, even asparagus. Flavours are often described as grassy or nettley.

sec – Dry wine style; French.

secco – Dry wine style; Italian.

Semillon – White grape variety originally of Bordeaux, where it is blended with Sauvignon Blanc to make fresh dry whites and, when harvested very late in the season, the ambrosial sweet whites of Barsac, Sauternes and other *appellations*. Even in the driest wines, the grape can be recognised from its honeyed, sweet-pineapple, even banana-like aromas. Now widely planted in Australia and Latin America, and frequently blended with **Chardonnay** to make interesting dry whites.

sherry – The great aperitif wine of Spain, centred on the Andalusian city of Jerez (from which the name sherry is an English corruption). There is a lot of sherry-style wine in the world, but only the authentic wine from Jerez and the neighbouring producing towns of Puerta de Santa Maria and Sanlucar de Barrameida may label their wines as such. The Spanish drink real sherry – very dry and fresh, pale in colour and served well chilled – called fino and manzanilla, and darker but naturally dry variations called amontillado, palo cortado and oloroso. The stuff sold under the big brand names for the British market are sweetened, coloured commercial yuck for putting in trifles or sideboard decanters to gather dust. The sherries recommended in this book are all real wines, made the way the Spanish like them. *See also* **fino** and **manzanilla.**

Shiraz – Australian name for the **Syrah** grape.

Spätlese – *see* **QmP.**

spritzy – Very slightly sparkling, as in the once-fashionable white *vinho verde* ('green wine') wines of the Minho Valley in northern Portugal. Faulty wines that have refermented a little after bottling may have a similar but thoroughly undesirable prickle.

Spumante – Sparkling wine of Italy.

stony – Wine-tasting term for keenly dry white wines. It's meant to indicate a wine of purity and real quality, with just the right match of fruit and acidity.

structured – Some wines are good in that the three stages of the flavours they impart in the mouth – the 'attack' or first impression, the middle palate as you 'chew' the wine, and the aftertaste – are very distinctly identifiable. Such a wine has structure.

summer fruit – Wine-tasting term intended to convey a smell or taste of soft fruits such as strawberries and raspberries – without having to commit too specifically to which.

Syrah – The noble grape of the Rhône Valley, France. Makes very dark, dense wine characterised by peppery, tarry aromas. Now planted all over southern France and farther afield. In Australia, where it makes wines ranging from disagreeably jam-like plonks to wonderfully rich and silky keeping wines, it is known as the **Shiraz.**

Tafelwein – Table wine, German. The humblest quality designation – doesn't usually bode very well.

tank method – Bulk-production process for sparkling wines. Base wine undergoes secondary fermentation in a large, sealed vat rather than in individual closed bottles. Also known as the Charmat method after the name of the inventor of the process.

tannin – One of those irritating words in the wine vocabulary. Tannin gets into red wine from the skins of black grapes. Well known as the film-forming, teeth-coating component in tea, tannin is a natural compound occurring in grape skins and acts as a natural preservative in wine. Its noticeable presence in wine is regarded as a good thing. It gives young everyday reds their dryness, firmness of flavour and **grip.** And it helps high-quality reds to retain their lively fruitiness for many years. A grand Bordeaux red when first made, for example, will have purply-sweet, rich fruit and mouth-puckering tannin, but after ten years or so this will have evolved into a delectably fruity, mature wine in which the formerly parching effects of the tannin have receded almost completely, leaving the shade of 'residual tannin' that marks out a great wine approaching maturity.

tears – The wine left clinging to the inside of the glass after the wine has been swirled round the glass then left to subside. *See also* **weight.**

Tempranillo – The great black grape of Spain. Along with Garnacha (Grenache in France) it makes all red **Rioja** and **Navarra** wines. Now grown in several regions of Spain and increasingly in South America.

tinto – On Spanish labels indicates a deeply coloured red wine. **Clarete** means a paler colour. Also Portuguese.

Toro – Quality wine region east of Zamora, Spain.

Torrontes – White grape variety of Argentina. Makes soft, dry wines often with delicious grapey-spicy aroma, similar in style to the classic dry **Muscat** wines of Alsace, but at more accessible prices.

trocken – 'Dry' German wine. Probably very dry indeed.

undercurrent – A whimsical personal note in which a wine is found to have a dominating flavour-association – say, blackcurrants – with something else detected on the aftertaste – say, strawberries.

Valdepeñas – An island of quality production amidst the ocean of mediocrity that is Spain's La Mancha region – where most of the grapes are grown for distilling into the head-banging brandies of Jerez. Valdepeñas reds are made from a grape they call the Cencibel – which turns out to be a very close relation of the Tempranillo grape that is the mainstay of the fine but expensive red wines of **Rioja**. Again like Rioja, Valdepeñas wines are matured in oak casks to give them a vanilla-rich smoothness. Among bargain reds, Valdepeñas is a name to look out for.

Valpolicella – Red wine of Verona, Italy. Good examples have ripe, cherry fruit and a pleasingly dry finish. Unfortunately, there are many bad examples of Valpolicella. Shop with circumspection.

varietal – A varietal wine is one named after the grape variety (one or more) from which it is made. Nearly all everyday wines, worldwide, are now labelled in this way. It is salutary to contemplate that just 20 years ago, wines described thus were virtually unknown outside Germany and one or two quirky regions of France and Italy.

Vermentino – White grape variety principally of Italy, especially Sardinia. Makes florally scented, soft dry whites.

Vin Délimité de Qualité Supérieur – Usually abbreviated to VDQS is a French wine-quality designation between Appellation Contrôlée and **vin de pays**. To qualify, the wine has to be from approved grape varieties grown in a defined zone. This designation is gradually disappearing.

vin de pays – 'Country wine' of France. The French map is divided up into more than 100 vin de pays regions. Wine in bottles labelled as such must be from grapes grown in the nominated zone or *département*. Some vin de pays areas are huge: the Vin de Pays d'Oc (named after the Languedoc region) covers much of the Midi and Provence. Plenty of wines bearing this humble appellation are of astoundingly high quality and certainly compete with New World counterparts for interest and value.

vin de table – The humblest official classification of French wine. Neither the region, grape varieties nor vintage need be stated on the label. The wine might not even be French. Don't expect too much from this kind of 'table wine'.

vinho de mesa – 'Table wine' of Portugal.

vino da tavola – The humblest official classification of Italian wine. Much ordinary plonk bears this designation, but the bizarre quirks of Italy's wine laws dictate that some of that country's finest wines are classed as mere *vino da tavola* (table wine). If an expensive Italian wine is labelled as such, it doesn't mean it will be a disappointment.

vino de mesa – 'Table wine' of Spain. Usually very ordinary.

Viognier – A grape variety once exclusive to the northern Rhône Valley in France where it makes a very chi-chi wine, Condrieu, usually costing up to £20. Now, the Viognier is grown more widely, in North and South America as well as elsewhere in France, and occasionally produces soft, marrowy whites that echo the grand style of Condrieu itself.

Viura – White grape variety of Rioja, Spain. Also widely grown elsewehere in Spain under the name Macabeo. Wines have a blossomy aroma and are dry, but sometimes soft at the expense of acidity.

weight – In an ideal world the weight of a wine is determined by the ripeness of the grapes from which it has been made. In some cases, it has to be said, that weight is determined by the quantity of sugar added during the production process. A good, genuine wine described as having weight is one in which there is plenty of alcohol and 'extract' – lots of colour and flavour from the grapes. Wine enthusiasts judge weight by swirling the wine in the glass and then examining the **'legs'** or **'tears'** left clinging to the inside of the glass after the contents have subsided. If these are dense and glyceriny and descend very slowly, the wine is deemed to have weight. A very good thing in all honestly made wines.

Winzergenossenshaft – One of the many very lengthy and peculiar words regularly found on labels of German wines. This means a wine-making co-operative. Many excellent German wines are made by these co-ops.

INDEX

SAVE £5.00 any six wines

This coupon entitles the holder to a discount of £5 off the total purchase price of any six bottles of wine at standard retail price only when presented at any Sainsbury's or Savacentre store. Offer limited to one coupon per purchase and is valid from 22/11/99 to 31/12/00 inclusive. The coupon can only be redeemed by persons over the age of 18.

Sainsbury's

To the Branch Manager
Retain the coupon in your till and return in the usual way.

9 904510 005009

Ref: BWB2000

Sainsbury's

Ref: BWB2000

SAVE £1.00

Coupon for £1.00 off any own label
bottle of wine valid at branches of
Londis and can only be redeemed for the
product listed at normal retail price.
No further discounts apply. Coupons valid
from 22/11/99 to 31/12/00 inclusive.
The coupon can only be redeemed by
persons over the age of 18.

To the Retailer
Retain the coupon in your till and return
in the usual way.

LONDIS
◆◆◆

Ref: BWB2000

SAVE 15% on any six wines

This coupon entitles the holder to 15% discount
on any six bottles of wine at standard retail price
when presented at any Wine Rack store. Offer
limited to one purchase per coupon and is valid
from 22/11/99 to 31/12/00 inclusive. No further
discounts apply, including shareholders or
loyalty discounts.
Offer available to persons aged 18 or over.
Coupon No. 659.

Wine Rack

To the Retailer
This voucher is redeemable as other voucher procedures
and should be collected and returned at the end of each week
in the normal manner.

Ref: BWB2000

SAVE 20% on any twelve wines

This coupon entitles the holder to 20% discount
on any twelve bottles of wine at standard retail
price when presented at any Bottoms Up store.
Offer limited to one purchase per coupon and is
valid from 22/11/99 to 31/12/00 inclusive. No
further discounts apply, including shareholders
or loyalty discounts. Offer available to persons
aged 18 or over. Coupon No. 659.

BOTTOMS·UP

To the Retailer
This voucher is redeemable as other voucher procedures
and should be collected and returned at the end of each week
in the normal manner.

Ref: BWB2000

LONDIS

Wine Rack

BOTTOMS·UP

SAVE 10% on any three wines

This coupon entitles the holder to 10% discount on any three bottles of wine at standard retail price when presented at any Threshers store. Offer limited to one purchase per coupon and is valid from 22/11/99 to 31/12/00 inclusive. No further discounts apply, including shareholders or loyalty discounts. Offer available to persons aged 18 or over. Coupon No. 659.

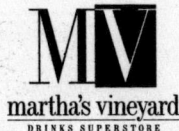

THRESHER WINE SHOP

To the Retailer
This voucher is redeemable as other voucher procedures and should be collected and returned at the end of each week in the normal manner.

Ref: BWB2000

SAVE 10% on any three wines

This coupon entitles the holder to 10% discount on any three bottles of wine at standard retail price when presented at any Victoria Wine store. Offer limited to one purchase per coupon and is valid from 22/11/99 to 31/12/00 inclusive. No further discounts apply, including shareholders or loyalty discounts. Offer available to persons aged 18 or over. Coupon No. 659.

VICTORIA WINE

To the Retailer
This voucher is redeemable as other voucher procedures and should be collected and returned at the end of each week in the normal manner.

Ref: BWB2000

SAVE 15% on any six wines

This coupon entitles the holder to 15% discount on any six bottles of wine at standard retail price when presented at any Martha's Vineyard store. Offer limited to one purchase per coupon and is valid from 22/11/99 to 31/12/00 inclusive. No further discounts apply, including shareholders or loyalty discounts. Offer available to persons aged 18 or over. Coupon No. 659.

MV
martha's vineyard
DRINKS SUPERSTORE

To the Retailer
This voucher is redeemable as other voucher procedures and should be collected and returned at the end of each week in the normal manner.

Ref: BWB2000

THRESHER WINE SHOP

Ref: BWB2000

VICTORIA WINE

Ref: BWB2000

MV

martha's vineyard

DRINKS SUPERSTORE

Ref: BWB2000